MANTLE CONVECTION FOR GEOLOGISTS

Mantle convection is the fundamental agent driving many of the geological features observed at the Earth's surface, including plate tectonics and hotspot volcanism. Yet many geologists have an incomplete understanding of the process, and misconceptions abound about how it relates to surface processes.

This book encourages Earth scientists to broaden their understanding of the physics and fluid dynamics of mantle convection by explaining what it is, how it works, and how to quantify it in simple terms. It assumes no specialist background: mechanisms are explained simply, and the required basic physics is fully reviewed and explained with minimal mathematics. The distinctive forms that convection takes in the Earth's mantle are described within the context of tectonic plates and mantle plumes, and implications are explored for geochemistry and tectonic evolution. Common misconceptions and controversies are addressed, providing a straightforward, but rigorous, explanation of this key process for students and researchers across a variety of geoscience disciplines.

DR GEOFFREY DAVIES is a Senior Fellow in the Research School of Earth Sciences at the Australian National University. He is an internationally honoured geophysicist who has been at the forefront of mantle convection studies for over three decades. He is also the author of the successful graduate textbook *Dynamic Earth: Plates, Plumes and Mantle Convection* (Cambridge University Press, 1999), and over 100 scientific papers. He has been at the forefront of attempts to reconcile mantle geochemistry with mantle dynamics, and in exploring the implications for the thermal and tectonic evolution of the Earth. Dr Davies was elected Fellow of the American Geophysical Union in 1990, and awarded the inaugural Augustus Love medal in geodynamics in 2005 by the European Geosciences Union.

MANTLE CONVECTION FOR GEOLOGISTS

GEOFFREY F. DAVIES

CAMBRIDGE UNIVERSITY PRESS
Cambridge, New York, Melbourne, Madrid, Cape Town, Singapore,
São Paulo, Delhi, Dubai, Tokyo, Mexico City

Cambridge University Press
The Edinburgh Building, Cambridge CB2 8RU, UK

Published in the United States of America by Cambridge University Press, New York

www.cambridge.org
Information on this title: www.cambridge.org/9780521198004

© Geoffrey F. Davies 2011

This publication is in copyright. Subject to statutory exception
and to the provisions of relevant collective licensing agreements,
no reproduction of any part may take place without the written
permission of Cambridge University Press.

First published 2011

Printed in the United Kingdom at the University Press, Cambridge

A catalogue record for this publication is available from the British Library

Library of Congress Cataloging in Publication data
Davies, Geoffrey F. (Geoffrey Frederick)
Mantle convection for geologists / Geoffrey F. Davies.
p. cm.
ISBN 978-0-521-19800-4 (hardback)
1. Plate tectonics. 2. Plumes (Fluid dynamics) 3. Earth – Mantle.
4. Heat – Convection. I. Title.
QE511.4.D376 2011
551.1'16 – dc22   2010041960

ISBN 978-0-521-19800-4 Hardback

Cambridge University Press has no responsibility for the persistence or
accuracy of URLs for external or third-party internet websites referred to
in this publication, and does not guarantee that any content on such
websites is, or will remain, accurate or appropriate.

Contents

1	Introduction	page 1
2	Context	4
	2.1 Crust, mantle, core	4
	2.2 Lithosphere versus crust	6
	2.3 Topography	8
	2.4 Heat flux	12
3	Why moving plates?	13
	3.1 The lead-up	13
	3.2 Wilson, plumes and plates	15
	3.3 Evidence for motion – magnetism	19
	3.4 Evidence for motion – seismology	22
	3.5 Evidence for motion – sediments	24
4	Solid, yielding mantle	26
	4.1 Viscosity	27
	4.2 Viscosity of the mantle	30
	4.3 Dependence of viscosity on temperature	35
	4.4 Inevitable convection	36
5	Convection	38
	5.1 Thermal expansion	40
	5.2 Buoyancy	41
	5.3 Plate velocity – simple mechanical version	42
	5.4 Heat conduction	45
	5.5 Plate velocity – thermo-viscous version	50
	5.6 The Rayleigh number and other fluid-dynamical beasts	52
6	The plate mode of convection	56
	6.1 The strong lithosphere	56
	6.2 The role of the lithosphere in mantle convection	59
	6.3 Heat transport – the plates are mantle convection	63

	6.4	The geography of topography	66
	6.5	The geography of heat flow	69
	6.6	Numerical model of the plate mode	69
	6.7	Summary of the plate mode	71
7		The plume mode of convection	73
	7.1	Inferring plumes from surface observations	75
	7.2	Hotspot swells, plume flows and eruption rates	78
	7.3	The dynamics and form of mantle plumes	84
	7.4	Plume heads and flood basalt eruptions	95
	7.5	Irregular volcanism and thermochemical plumes	99
	7.6	Summary of the plume mode	101
8		Perspective	104
	8.1	Separate but interacting	104
	8.2	Common misconceptions	107
	8.3	Layered convection?	112
	8.4	Other modes and causes	116
	8.5	Pursuing implications	122
9		Evolution and tectonics	124
	9.1	Parametrised thermal evolution	125
	9.2	Numerical thermal evolution	130
	9.3	The heat source puzzle	132
	9.4	Compositional buoyancy	136
	9.5	Intermittent plate tectonics	147
	9.6	Implications for tectonics	147
10		Mantle chemical evolution	154
	10.1	Trace element and isotope observations	155
	10.2	Global budgets	160
	10.3	Incompatible trace elements in the mantle	162
	10.4	Mantle heterogeneity	164
	10.5	Melting in a heterogeneous mantle	171
	10.6	Previous estimates of trace element abundance	176
	10.7	Dynamical modelling of refractory incompatible elements	184
	10.8	Resolving the noble gas enigma	193
	10.9	Assessment of mantle chemistry	205
11		Assimilating mantle convection into geology	206
		Appendix A Exponential growth and decay	208
	A.1	Exponential solution	208
	A.2	Post-glacial rebound	209
	A.3	Rayleigh–Taylor instability	210

Appendix B	*Thermal evolution details*	211
B.1	Heat generation	211
B.2	Parametrised thermal evolution model	211
B.3	Numerical thermal evolution model	213
B.4	Basalt tracers	213
B.5	Models with basalt tracers	214
Appendix C	*Chemical evolution details*	216
C.1	Fraction of primitive mantle	216
C.2	Fraction of crust in the mantle	216
References		218
Index		230

1
Introduction

Mantle convection is the fundamental agent driving most geology, yet many geologists still have only vague ideas about what mantle convection is, how it works and how it might inform their specialty. Because it is so fundamental, the better every geologist understands mantle convection, the better scientist he or she is likely to be. Of course, not everything is affected by mantle convection, but only by being well informed will a geologist recognise when it is relevant, and what that relevance is.

Misconceptions about mantle convection also seem still to be quite widespread. Some aspects of mantle convection are debated. Much of that debate concerns refinements, so the debate is quite legitimate, but some of the debate is based on misconceptions or incomplete understanding of current theories or observations. The latter debate is not productive. This is not to claim that alternative versions are inconceivable, but just to note that debaters need to be informed about the theories they wish to challenge if they are to make useful contributions.

For these reasons it seems worthwhile to offer an account of our current understanding of mantle convection in terms that are reasonably accessible to most geologists. That means the account should be fairly short, and there should be little mathematics beyond basic algebra and arithmetic. Nor should a strong grasp of physics be assumed, and such physics as is required (notably heat conduction and viscous fluid flow) should be explained in simple and reasonably familiar terms. These are the constraints I have set in writing this book.

Actually I thought I had already done this in *Dynamic Earth* [1]. The essential arguments are presented there in fairly simple terms, and mathematical or detailed sections are clearly flagged and can be skipped. However, that book is quite long, it is not cheap, and the appearance of equations is doubtless intimidating. Also, *Dynamic Earth* is probably regarded as *geophysics*, and few geologists might therefore bother to peruse it. So, it seems *Dynamic Earth* has not accomplished my present purpose and a separate account is required that is explicitly directed to geologists.

I use the term *geologist* in the broad sense of anyone who studies geological processes. This includes not only field geologists of various kinds, such as structural geologists, but also petrologists, geochronologists, geochemists, ore geologists, sedimentologists, palaeontologists and so on. It also includes geophysicists. For example, seismologists are often so specialised in the intricacies of their discipline that they have little understanding of other aspects of geophysics.

So, returning to the relevance of mantle convection to geology, there are two energy sources driving geological processes. Solar energy drives surficial processes related to the weather and life, such as weathering, erosion and sediment transport. All other geological processes are driven by the Earth's internal heat. These include mountain building, or more generally tectonics, magma generation, water flows in the deeper crust, metamorphism and much mineral deposition.

Tectonics, meaning the movements of the crust that result in mountains, rifts, faults, folds and so on, is driven by the Earth's internal heat and is obviously fundamental to a large proportion of geological processes. The connection between heat and tectonics is through convection: convection in the mantle is driven by the Earth's internal heat, and it generates the movements that manifest as tectonics. Plate tectonics is the primary agent of tectonics, and its existence is now widely accepted. Volcanic hotspots (by which I mean surface features like the Hawaiian and Icelandic volcanic centres) and their relatives, the flood basalts, are a secondary tectonic agent, and their existence and basic features are also widely recognised. There may be other tectonic agents, but evidently they are minor.

This much is well known and widely understood. However, the causes of plate motions and volcanic hotspots still do not seem to be clearly understood in the broad geological community. They are usually understood to involve mantle convection, but the relationship of plates and plumes to that convection seems to be understood only vaguely, and often with some basic misconceptions – or so it seems from my encounters with non-specialist colleagues and students.

To many non-specialists, mantle convection seems to be something rather mysterious that happens 'down there'. Its relationship with plates is not very clear. The idea of mantle plumes is hotly disputed by a few, and plumes are not uncommonly regarded as arbitrarily adjustable to fit circumstances and therefore not real science. Other confusions may exist, such as between crust and lithosphere, and misconceptions are not uncommon, such as that there are warmer upwellings rising under mid-ocean ridges, that 'plume tectonics' is an alternative to plate tectonics, or that mantle plumes are molten.

Yet there are now straightforward and well-quantified physical theories that account for the main features of plate movements, volcanic hotspots and flood basalts. Furthermore you don't have to be a rocket scientist to understand the essence of the physical theories – they can be understood fairly readily without a lot of mathematics or a PhD in physics.

The main purpose of this book is to focus on the questions of why plates move and what causes volcanic hotspots and flood basalts, and to present answers in a way that any geologist should be able to understand. In the view presented here, moving plates and volcanic hotspots are manifestations of convection in the Earth's mantle. Convection can be understood on the basis of two kinds of basic physical process (heat conduction and viscous fluid flow), which are explained through simple examples. Putting these two physical processes together allows one, for example, to calculate fairly simply how fast plates should move, and to get an answer that is close to what we observe. In the process, the relationship between plates and convection becomes clear, the likelihood of narrow, warm, columns rising through the mantle becomes evident, and other major features of Earth's tectonic system fall into place.

The book then goes on to look at two kinds of implication. The first is the evolution of the mantle, and how that might relate to tectonic evolution. The second is how geochemistry might fit into the physical and dynamical picture. This has been a vexed issue for some time, but refractory trace elements have by now been plausibly and quantitatively incorporated, and the noble gases might now also be finding a place. This is important because the geochemistry provides information not available just from the physics.

I should be clear that in offering 'answers' I don't mean 'the truth', I mean theories well based in physics that can account for many of the main features we observe. As always in science, this does not mean that better theories might not emerge, nor that alternative theories do not exist. Choosing between alternative theories is not entirely a rational process, it also involves judgements, and this book reflects my own judgements. In my experience, some of the extant criticisms of the theories presented here reflect a lack of clear understanding of what the theories actually are. So there are two roles this presentation can play: to inform the professionally curious, and to focus debates more constructively on real issues instead of misunderstandings.

2
Context

> Basic concepts and primary observations. Defining the crust, mantle and core. Distinguishing crust from lithosphere, continents from ocean basins. The distribution of topography and heat flux over the sea floor.

Mantle convection occurs, remarkably enough, in the Earth's mantle. It is affected by the crust, and part of the lithosphere plays a major role. There are peculiarities near the boundary of the mantle with the core that may significantly affect mantle convection, and that certainly tell us some important things about mantle convection. To discuss our subject sensibly, we had better be clear what all these terms refer to: mantle, crust, core, lithosphere and so on. That is one thing this chapter is about. There are also important constraints on mantle convection to be had from the form of the Earth's topography, and from the geographic variation of heat flow from the Earth's interior. These will also be summarised.

2.1 Crust, mantle, core

The major division of the Earth's interior is into crust, mantle and core. The boundaries between these regions were detected seismologically, in other words using the internal elastic waves generated by earthquakes, which are detected as they emerge at the Earth's surface. The variation of seismic velocities, and density, with depth in the Earth is shown in Figure 2.1. The boundary between the mantle and the core is at a depth of about 2900 km, where the seismic velocities drop, the shear velocity is zero and the density jumps.

The fact that the shear velocity is zero in the core indicates that it is liquid, except for a smaller region at the centre, the inner core, which is solid. The high density of the core is consistent with it being made mostly of iron, with some nickel and

2.1 Crust, mantle, core

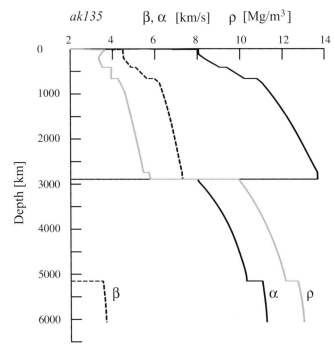

Figure 2.1. Variation of properties with depth in the Earth, defining crust, mantle and core. Profiles are of the seismic compressional velocity, α (solid black), shear velocity, β (dashed black), and density, ρ (solid grey). Curves from the model ak135. Figure courtesy of Kennett *et al.* [2, 3].

some lighter elements. The inference of iron and nickel comes from meteorites, some classes of which are made of an iron–nickel alloy.

The boundary between the crust and the mantle is barely discernible in Figure 2.1, because the crust is so thin on this scale, as will be discussed below. There are two jumps within the mantle, at depths of 400 km and 660 km. These define the *transition zone*, and are the locations of pressure-induced phase transformations, where the mantle minerals collapse into denser crystal structures. The transition zone may have played a large role in determining the form of mantle convection early in Earth's history. For many years there was also a major debate about whether the 660 km jump separated convection in the upper mantle from convection in the lower mantle, but there is strong evidence now that convection passes through the transition zone in the modern Earth. This will be discussed in later chapters.

Also visible in Figure 2.1 are changes in the bottom 200–300 km of the mantle. These changes are not well resolved in this model, but other seismic studies have clearly identified changes in seismic velocity and in some places discontinuities. This zone is known as the D″ region, terminology left over from early studies of

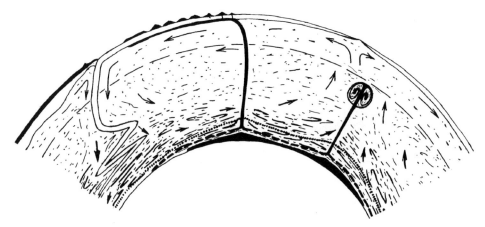

Figure 2.2. Sketch of the mantle, with crust and lithosphere roughly to scale. The core is defined by the bottom boundary of the mantle. From Davies [4]. Copyright by Elsevier Science. Reprinted with permission.

Earth's interior. There is now good evidence that it is due to a combination of a change in composition and one or perhaps two pressure-induced phase changes.

Based on this information, and other sources that we will encounter through the book, Figure 2.2 is a sketched cross-section of the crust and mantle, roughly to scale. The continental crust and some thicker parts of the oceanic crust show as black at the surface, though their thickness is exaggerated. The lithosphere, defined below, is outlined by the thin black lines. The lower boundary of the mantle outlines the extent of the core. Within the mantle, the 660 km discontinuity is marked by the long-dashed line. The D″ region is outlined by dashed and dotted lines at the bottom of the mantle. Other features in this sketch will be explained later in the book.

2.2 Lithosphere versus crust

The oceanic lithosphere plays a key role in the conception to be developed in this book. The role of continental lithosphere seems to be much less central, largely because continental crust is different from oceanic crust. The distinctions between crust and lithosphere, between continental and oceanic lithosphere, and between continental and oceanic crust thus need to be clear, otherwise the discussion of mantle convection will be confused. The distinctions are illustrated in cartoon form in Figure 2.3, which is not to scale.

The continental crust is commonly about 35–40 km thick. This was first determined in 1909 when Mohorovičić identified the seismic discontinuity named after him [5], also known as the 'moho'. The thickness is larger under mountain ranges,

2.2 Lithosphere versus crust

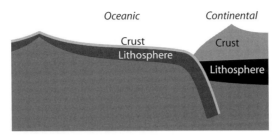

Figure 2.3. Distinctions between crust and lithosphere, and between their continental and oceanic forms.

reaching nearly double this value under the Himalayas. It is smaller in some regions where the crust has been horizontally extended. The average thickness seems to be determined by a long-term balance between horizontal shortening and erosion. This was first perceived by Hess [6], who argued that continental crust tends to be eroded down to sea level and pushed up by the plate tectonic forces that he was among the first to conceive.

The thickness of the oceanic crust was not determined until the 1950s and 1960s, using seismic surface waves and later ocean-going seismic profiling [7]. To most geologists' surprise, it turned out to be much thinner than continental crust, averaging about 7 km. It is thicker in some places where there is an 'oceanic plateau', and it is thinner in only a few places. Otherwise it is remarkably uniform in thickness.

Both kinds of crust were detected and defined seismically. Seismic waves travel more slowly in the crust than in the underlying mantle rocks. This contrast in wave velocity causes reflections and refractions, and these allow seismologists to infer the presence of interfaces or 'discontinuities' below the Earth's surface.

The difference in seismic velocity between crust and mantle implies a difference in composition. This inference is not a simple one, as there was debate for some time about whether the continental moho might be due to a phase change, in which the minerals comprising the rocks are squeezed into denser crystal structures due to the increase of pressure with increasing depth. Eventually detailed studies resolved the debate in favour of a change in composition. There was also a debate about the oceanic moho. Hess [6], for example, proposed that the oceanic crust was made of serpentinite, which is a hydrated form of the predominant upper-mantle mineral olivine. There are places where hydrated mantle is known, but drilling has established that most of the oceanic crust has a basaltic composition, different from the underlying mantle, whose dominant rock type is peridotite.

The concept of the lithosphere was established by early in the twentieth century. During the nineteenth century, geologists established evidence for continuing uplift and subsidence during geological history, as distinct from everything being frozen in place since early in Earth's history. This required the interior of the Earth to be deformable, though not necessarily liquid. Seismology established that the mantle is in fact not liquid, so the mantle was inferred to be a deformable solid. This history will be discussed in more detail in Chapter 4. Despite the evidence for continuing deformations, it was also evident that structures less than about 100 km in horizontal extent seemed to be supported without continuing deformation. These observations can be reconciled if the outer 100–200 km of the Earth is strong and usually not deforming, even on geological timescales.

This outer, strong layer became known as the *lithosphere*. In 1914 Barrell [8] proposed the term *asthenosphere* for the deformable region below the lithosphere. Thus the lithosphere is defined in terms of its *strength*. Since it is thicker than the crust, it must comprise the crust and the top part of the mantle.

The greater strength of the lithosphere was inferred to be due to lower temperatures near the Earth's surface, and this is confirmed in the modern picture, as we will see as we go along. The lower temperatures also cause seismic velocities to be higher, and modern techniques and instruments have allowed the lithosphere to be resolved seismically, although it is more subtle and harder to distinguish than the crust. In this way we have learnt that the oceanic lithosphere is up to around 100 km thick, though it is thinner near mid-ocean ridges. On the other hand, the continental lithosphere is rather variable, and over 200 km thick under older parts of the continental crust.

To summarise, the *lithosphere* is defined by its strength, which is sufficient to prevent it from deforming significantly on geological timescales. Its strength is inferred to be due to its lower temperature. It is up to about 100 km thick in oceanic regions, and from about 100 km to over 200 km thick in continental regions. The *crust* is defined by its lower seismic wave velocities. The lower velocities are inferred to be due to it having a different composition than the mantle underneath. In oceanic regions it is about 7 km thick and has a basaltic composition. Its density is about 2900 kg/m^3, in contrast to the upper mantle density of 3300 kg/m^3. In continental regions it averages 35–40 km thick. Its composition is quite variable, roughly from basaltic to granitic, and averaging to an intermediate rock type like andesite. Its density is also rather variable, and averages around 2700 kg/m^3. These properties are summarised in Table 2.1.

2.3 Topography

The Earth's topography has some striking features. To appreciate it fully, we need to see it without the oceans, as it is shown in Figure 2.4. In this view it is very clear that there are two predominant elevations of the Earth's surface, that of the continents and that of the ocean basins. This bimodal distribution of elevation is

Table 2.1. *Defining characteristics of crust and lithosphere.*

Characteristic	Continental	Oceanic
Crust		
Defining property	low seismic velocity	low seismic velocity
Reason	composition	composition
Average composition	andesitic	basaltic
Average thickness	35 to 40 km	7 km
Average density	2700 kg/m^3	2900 kg/m^3
Lithosphere		
Defining property	strength	strength
Reason	low temperature	low temperature
Thickness	100 to >200 km	0 to 100 km

a striking feature, not observed on any other body in the solar system. Hess [6] argued that it is due to the combined action of seafloor spreading, which sweeps continental material together, and subaerial erosion, which planes the continental surfaces down to sea level.

The next most prominent feature of the Earth's topography is the system of mid-ocean ridges, which form a continuous network within the ocean basins. These stand 2–3 km above the deeper parts of the ocean basins, which are 5–6 km below sea level. They will give us important information about mantle convection. Not very visible in this image are the deep ocean trenches, bordering the Pacific basin, Indonesia and a few other places. These extend to depths of around 10 km below sea level. Various plateaus and mountain belts are visible on the continents, and various plateaus and chains of seamounts are visible in the ocean basins. A broad swell in the sea floor is visible around Hawaii and the chain of seamounts extending northwest from Hawaii, in the mid-North Pacific. This, and a few features like it, will also tell us something important about mantle convection.

There is, in the seafloor topography, a surprising regularity that is not obvious just from a map like Figure 2.4. It is that seafloor depth correlates strongly with the age of the sea floor. More specifically, it correlates with the square root of the seafloor age. This is illustrated in Figure 2.5. Obviously there are deviations from the correlation, but overall the seafloor age is the main predictor of seafloor depth.

There is also some regularity in the deviations from the main correlation in Figure 2.5. The deviations are mainly positive, and they occur mainly on older sea floor. In other words, there is a tendency for older sea floor to be shallower than the correlation predicts. This tendency is not universal, however. For example, profiles 1 and 2 extend, with only minor deviations, to ages of 175 Ma and 100 Ma, respectively.

10 *Context*

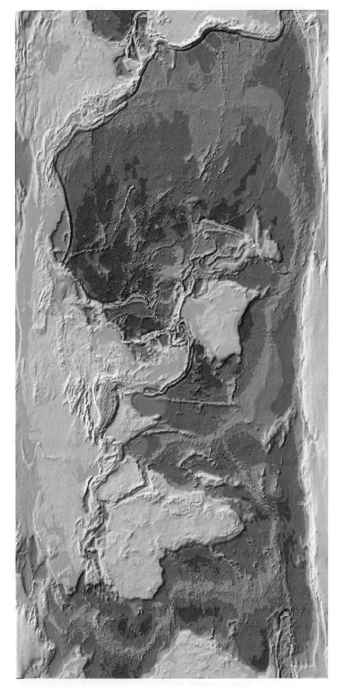

Figure 2.4. Topography of the Earth. The submarine breaks in the grey scale are at depths of 5400 m, 4200 m, 2000 m and 0 m. Shading of relief is superimposed, with a simulated illumination from the northeast. From the ETOPO5 data set from the US National Geophysical Data Center [9]. Image generated using *2DMap* software, courtesy of Jean Braun, Australian National University.

2.3 Topography

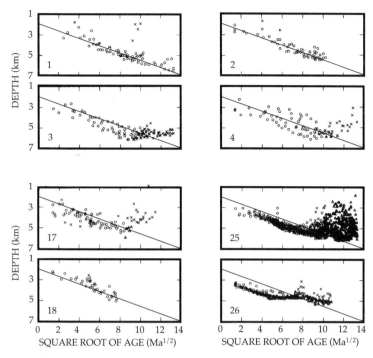

Figure 2.5. Seafloor depth versus square root of age for a selection of regions of the sea floor. The same reference line is shown in each plot. After Marty and Cazenave [10]. Copyright by Elsevier Science. Reprinted with permission.

There is a well-entrenched story that sea floor older than about 80 Ma approaches a uniform depth. This story is a myth. It originated when the data coverage of the sea floor was sparse, and it survives because people keep aggregating the data into a single global distribution, thus missing important geographical information that is evident in Figure 2.5. Further evidence for significant regional variation comes from variations in the depth of the crests of mid-ocean ridges. These variations are not clearly evident from Figure 2.4, although you can see that the ridge crest south of Australia is unusually deep, and the crest of the East Pacific Rise seems to be unusually shallow. Profiles 25 and 26 in Figure 2.5 also exhibit low depths of ridge crests. A more accurate description of the observations is that there are regional deviations from the main depth–age distribution that have an amplitude of about 1 km.

It is worth emphasising how remarkable the main depth–age correlation of the sea floor appears. Geologists working on the continental crust never see such regularity, because the continental crust has been so thoroughly and pervasively worked over by tectonic processes. Even for fluid dynamicists used to fluid convection, this is a remarkably regular surface, not usually seen in convection systems. We will see later that important insights can be gained from this simplicity.

12 *Context*

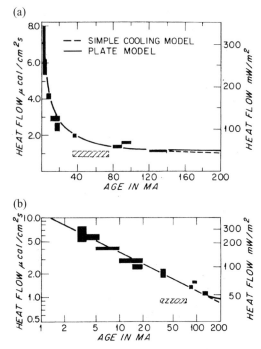

Figure 2.6. (a) Oceanic heat flux versus age of sea floor. The two curves are explained in the text. The shaded box represents observations that were superseded by the data at 40 Ma. (b) The same data and curves on a logarithmic plot, in which the 'simple cooling model' predicts a straight line with a slope of $-1/2$. From Sclater *et al.* [11]. Copyright by the American Geophysical Union.

2.4 Heat flux

The distribution of heat flowing out of the Earth's interior also carries important information. Again there is a correlation with age of sea floor (Figure 2.6(a)) and with the square root of the age (Figure 2.6(b)), although this time the correlation is inverse: the greater the age, the lower the heat flow. The dashed curve ('simple cooling model') is the best-fit line in which heat flux is inversely proportional to the square root of age. The solid curve ('plate model') is a model based on the false interpretation, just discussed, that seafloor depth approaches a constant value. The models of and insights to be gained from the heat flux data will be encountered later.

The concepts and observations summarised in this chapter will allow us to make sense of plate tectonics and mantle convection. Other evidence is also useful, and some has perhaps had more prominence, but what we have here is powerful, and sufficient to infer the main picture. Some of the additional evidence will be mentioned along the way.

3

Why moving plates?

> How were plates conceived of? How do we know they move? How were mantle plumes conceived?

The main story of the idea of continental drift and its ultimate transformation into the theory of plate tectonics has been well told (e.g. [12, 13]), and will not be recounted in detail here. However, the particular observations that led to the conception of moving plates are perhaps not so well known, as evidenced by the lack of recognition of the person who first conceived of plates. Also, the evidence that persuaded large segments of the geophysical and broader geological community of the reality of moving plates is worth repeating, so we know we are dealing with a theory well based in observations, and not just accepted on the authority of textbooks already a few decades old. (I say 'already' because I came into the Earth sciences in 1968, just after the view of the geophysical community had been transformed, and before many geologists had been persuaded, so of course to me it doesn't seem long ago.)

3.1 The lead-up

Alfred Wegener conceived of continental drift around 1912 [14] on the basis of the rough match of continental outlines across the Atlantic Ocean, which he was not the first to notice. He supported the concept with geological and palaeontological evidence. He refuted with sound physics the rival theory of land bridges that had risen and sunk, notably that there should be large and observable gravity anomalies associated with such changes. His arguments were opposed by people, notably Harold Jeffreys [15], who had little knowledge of the relevant properties of solids, and in particular apparently little awareness of good geological arguments at the time, which will be discussed in the next chapter, that the mantle must be deformable to accommodate slow vertical motions of the surface. Wegener died on the Greenland ice sheet in 1930, and his theory fell into disrepute in the anglophone world, with the notable exceptions of Arthur Holmes [16] in the UK, Reginald Daly [17] in the USA, Alex du Toit [18] in South Africa and Sam Carey [19] in Australia.

The idea of continental drift was revived in the late 1950s on the basis of geomagnetic observations of apparent polar wander [20], though it did not gain wide currency. Exploration of the sea floor after World War II led to discoveries of features that had no equivalents on land, such as the mid-ocean ridge system and fracture zones thousands of kilometres long. These, in turn, generated a ferment of ideas that might explain them, among which were an expanding Earth and continental drift [7]. Eventually the marine evidence led Hess and Dietz [6, 21] independently to the idea of seafloor spreading. In their conception there also had to be a complementary subduction of sea floor at ocean trenches, but the evidence was more circumstantial at trenches.

Mantle convection was very much a part of the exploratory thinking during this period, but it is interesting that in some ways this was a hindrance. For example, Heezen [22] traced the Mid-Atlantic Ridge south of Africa and into the Indian Ocean, and then perceived a problem. The ridges were inferred to be the sites of horizontal extension of the crust, because wherever they came on land they were extensional, such as in Iceland and around the Red Sea, and there is commonly a trough at their crests, which is consistent with the presence of an extensional graben. If both the Atlantic and Indian ridges are extending, then, to Heezen, Africa should be undergoing shortening, but there is no evidence for such shortening. Heezen's suggested resolution of this problem was that the Earth must be expanding.

There is an important lesson to be learned from looking at Heezen's logic. Heezen, like most geologists of the time, thought of convection in terms of 'cells', with an active upwelling on one side and an active downwelling on the other, as depicted in Figure 3.1(a). Figure 3.1(a) also includes schematic locations of continents and ridges. There is subduction on the western side of the Americas that can accommodate the extension of both the East Pacific Rise (EPR) and the Mid-Atlantic Ridge (MAR), so that part seems reasonable. However, there is no subduction between the Mid-Atlantic Ridge and the Mid-Indian Ridge (MIR). Therefore the extension at the ridges must be increasing the area between them. Therefore, according to Heezen, the Earth must be expanding.

The other possibility is that the ridges are moving relative to each other, that is, the MIR is moving away from the MAR. The reason this did not seem possible to Heezen is that he presumed that there would be hot, active upwellings under the ridges, and if the ridge moved it would move away from the upwelling and cease to spread. The resolution of this puzzle is that convection does not have to take

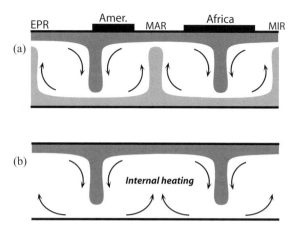

Figure 3.1. (a) Heezen's conception of the convection cells underlying the mid-ocean ridge system. EPR, East Pacific Rise; MAR, Mid-Atlantic Ridge; MIR, Mid-Indian Ridge. (b) Internally heated convection has active downwellings, but no active upwellings. Ridges are therefore free to move.

the form of the textbook-style 'cells' sketched in Figure 3.1(a). This form occurs only when the fluid is heated from below, thus generating hot upwellings, and only if the heating is not too strong. If the heating is stronger, upwellings and downwellings tend to move about. More importantly here, if the fluid is heated internally, which is a more accurate picture of the mantle case, there are no hot upwellings (Figure 3.1(b)). There must be upwelling between the active downwellings, but it is simply the fluid being passively displaced by the downwellings. If there are no hot, active upwellings, then a spreading centre can be located anywhere, or can move about. We will look at this again in a later chapter.

3.2 Wilson, plumes and plates

Tuzo Wilson cut through the prevailing confusion when he followed the logic of surface features, without worrying about what might be happening underneath [23]. The seafloor spreading papers of Hess and Dietz led Wilson to change his ideas. Until then he had been a 'fixist', believing that continents did not move, but those papers convinced him that they must move. He set about finding more evidence. This led him to not one but two seminal ideas, both central to understanding mantle convection. The first idea did involve thinking about what is below the surface, but the second did not.

Wilson first thought of using ocean islands as probes of the sea floor, reasoning that islands should be progressively older at greater distances from spreading centres [24]. The idea was good, but the data were scattered and somewhat misleading, since some islands include fragments of continental crust, and other ages were not representative of the main phase of island formation. Also, even an accurate age for the main phase of building an island gives only a lower bound on the age of the sea floor upon which it is built. This is because the sea floor will have the same age as the island if the island formed at a spreading centre, but it will be older if the island formed away from the spreading centre.

Despite the limited data available at the time, this phase of Wilson's work led him to his first seminal idea, which was a mechanism to explain age progressions in island chains. Darwin had been the first to recognise an apparent age sequence among Pacific islands, from an island with fringing reef, through an island with barrier reef, to an atoll. However, it was Dana who first observed such a sequence among adjacent islands, and he extended the sequence to include an initial volcanically active island ([7] p. 195). Wilson was aware that the Hawaiian island chain was a classic example of such a sequence, though the ages of the islands were known only qualitatively. With the idea of sea floor moving sideways, he realised that Dana's inferred age sequence for the Hawaiian islands could be produced if there was a (relatively) stationary source of volcanism deep in the mantle that had generated the islands successively as the sea floor passed over [25]. He conjectured that this 'hotspot' source might be located near the slowly moving centre of a convection 'cell'. This particular mechanism was soon superseded by the idea of a mantle plume, which we will develop later in the book, but the idea of a relatively stationary deep-mantle source of volcanism is common to both hypotheses. Accurate dating of the islands soon provided strong support for the idea [26, 27].

Wilson was a physicist turned geologist with very broad interests in geology, but he had a particular interest in large transcurrent faults. His recognition of the North American equivalent (the Cabot fault) of Scotland's Great Glen fault fuelled his interest in continental drift. In 1963, the same year his work on island chains was published, he also published a very wide-ranging discussion of continental drift [28]. In this, it is clear he had a comprehensive grasp not only of a large number of geological observations but also of the arguments from isostasy, post-glacial rebound, materials science and gravity observations over ocean trenches that the mantle is deformable and plausibly undergoing convection, as will be discussed in the next chapter.

Wilson was also puzzled by the great fracture zones that were being discovered on the ocean floor, because they seemed to be transcurrent faults of large displacement, but they stopped at the continental margin, with no equivalent expression on the adjacent continent. His clinching insight was his recognition of the way these great faults can connect consistently with mid-ocean ridges or with 'mountains' (meaning island arcs or subduction zones) if pieces of the crust are moving relative to each other as rigid blocks without having to conserve crust locally. This led him

3.2 Wilson, plumes and plates

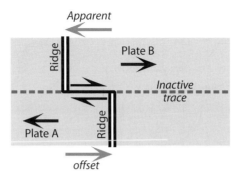

Figure 3.2. Sketch of one kind of transform fault, the kind that joins two spreading ridge segments.

to define *transform faults*, for which he is perhaps best known. Figure 3.2 illustrates one kind of transform fault. Wilson recognised that, if the plates on either side of the ridge segments are separating, then the sense of motion on the fault joining the ridge segments will be right lateral, whereas the apparent offset of the ridge crest is left lateral. At the point where the fault meets a ridge segment, the right lateral motion is *transformed* into horizontal extension along the ridge segment. A transform fault can also connect with a subduction zone, so a transform fault is a transcurrent (strike-slip) fault at whose ends the strike-slip motion is transformed into horizontal extension or shortening.

Wilson's famous paper of 1965, 'A new class of faults and their bearing on continental drift' [23] is perhaps still not as well appreciated as it should be. In this paper Wilson not only defined transform faults, but also went on for the first time to describe a globally connected set of mobile belts that divide the Earth's surface into a giant jigsaw of moving pieces, which he called *plates*. In other words, he conceived for the first time of plate tectonics. It is clear from the second part of his title, and from his comprehensive discussion of continental drift of two years earlier [28], that he knew he had solved the puzzle of continental drift in the course of conceiving a grander and more comprehensive theory. It is worth quoting the opening of the paper [23].

Many geologists [29] have maintained that movements of the Earth's crust are concentrated in mobile belts, which may take the form of mountains, mid-ocean ridges or major faults with large horizontal movements. These features and the seismic activity along them often appear to end abruptly, which is puzzling. The problem has been difficult to investigate because most terminations lie in ocean basins.

This article suggests that these features are not isolated, that few come to dead ends, but that they are connected into a continuous network of mobile belts about the Earth which divide the surface into several large rigid plates.

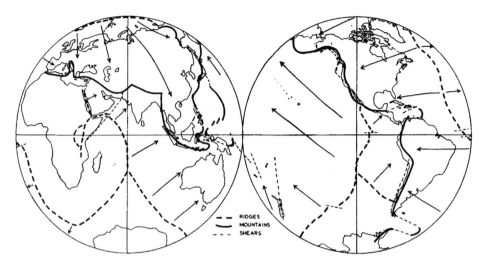

Figure 3.3. Wilson's 1965 sketch map of the plates [23]. Reprinted from *Nature* with permission. Copyright Macmillan Magazines Ltd.

In these four sentences Wilson defined the problem and presented its solution with simple clarity. His sketch map (Figure 3.3) gave the world its first view of the tectonic plates. It is only a sketch map, but the concepts are clear and precise.

Wilson was thinking as a structural geologist, and that was crucial. He envisaged rigid blocks bounded by three types of boundary that correspond to the three standard fault types: strike-slip (transform fault), normal (ridge) and reverse (subduction zone). Conceptually, he narrowed the old notion of mobile belts down to sharp boundaries, and he explicitly adopted the long-standing implication of that old term, that there is little deformation outside the mobile belts, taking it conceptually to the limit of proposing that there is *no* deformation. He was explicit in the fourth sentence, quoted above, that the plates are 'rigid'. The point is explicit also within the paper: 'These proposals owe much to the ideas of S. W. Carey, but differ in that I suggest that the plates between the mobile belts are not readily deformed except at their edges.'

By looking and thinking as a structural geologist, Wilson was able to see the plates in all their simplicity. A unique and crucial feature of mantle convection, as distinct from other forms of convection, is that the convecting material behaves sometimes as a viscous fluid and sometimes as a brittle solid, as we will see in later chapters. This was a major source of the confusion in earlier attempts to formulate and relate ideas about continental drift, seafloor spreading and mantle convection. This can be seen, for example, by contrasting Arthur Holmes' earlier concept of a new ocean in which there is broad deformation across the sea floor, reflecting the behaviour of a viscous fluid, with Hess' and Dietz' narrow spreading centres, and with the angular, segmented geometry of mid-ocean ridges. By focusing on the motions that can be discerned at the surface, Wilson recognised the behaviour of a brittle solid, and successfully defined plate tectonics in those terms.

There was significant supporting evidence already when Wilson published his theory. It was known that earthquakes do not occur on the fracture zones beyond where they connect with offset ridge crests, but only on the segment between ridge crests. This is consistent with Wilson's description of how the two plates in Figure 3.2 pull apart. The transform fault between the ridge segments is active. Beyond the ridge segments, the two sides of the fracture zone are moving together, so there is no relative motion and no earthquakes would be expected.

It took a couple of years for Wilson's idea to catch on, but when it did, other strong evidence was soon found supporting his theory. I will note three kinds of evidence that are particularly direct, and speak to different sections of the Earth science community.

3.3 Evidence for motion – magnetism

In about 1960, studies of palaeomagnetism began to focus strongly on the question of whether the Earth's magnetic field has changed through time. Specifically, has it reversed polarity? Matuyama, in 1929, had studied the magnetisation through a sequence of lava flows erupted by a Japanese volcano. He found that the younger flows near the top were magnetised parallel to the present Earth's magnetic field lines, but that the older flows near the bottom of the sequence were magnetised in the opposite direction. During the 1950s the question of whether this was due to reversal of the Earth's field or to a peculiar response of some rocks was vigorously debated. It seemed that there may have been many reversals of the Earth's field, but this was difficult to demonstrate convincingly. From 1963, two groups in particular used a combination of magnetisation measurements and potassium–argon (K–Ar) dating to try to resolve the question and to establish a chronology of reversals. These groups were at the US Geological Survey in Menlo Park, California, and at the Australian National University in Canberra. They found that the ages of normally and reversely magnetised rocks correlated around the world, which support the idea that the Earth's field had indeed reversed. By about 1969, the time sequence of reversals was established with some detail to an age of about 4.5 Ma, beyond which the K–Ar dating method did not have sufficient accuracy.

Meanwhile, Ron Mason, of Imperial College, London, and the Scripps Institute of Oceanography in California, was trying to identify magnetic reversals in oceanic sedimentary sequences. Because of this work, but still almost by chance, a magnetometer was towed behind a ship doing a detailed bathymetric survey off the west coast of the USA. From this magnetic survey there emerged a striking and puzzling pattern of variations in magnetic intensity: alternating strips of the sea floor had stronger and weaker magnetic field strengths. The pattern was parallel to the local fabric of seafloor topography, and later was found to be offset by fracture zones, by about 1000 km in the case of the Mendocino fracture zone. It was presumed that the pattern might be explained by strips of sea floor with differing magnetisations, but its origin was obscure. In retrospect it was unfortunate that in this area the ocean spreading centre at which the sea floor formed no longer exists, and so there was no obvious association with mid-ocean ridges.

In subsequent surveys, elsewhere, it was found that ridge crests have a positive magnetic anomaly (meaning merely that the field strength is greater than average), which some people presumed to indicate 'normal' (i.e. not reversed) magnetisation. Beyond this, on either side of the ridge crest, there was a negative (i.e. weaker than average) anomaly. In 1963 Fred Vine, then a graduate student at Cambridge University, was analysing the results of one such survey over the Carlsberg Ridge in the Indian Ocean. He noticed that the seamounts near the ridge crest were reversely magnetised. This is easier to infer for seamounts, because they are more like point sources and produce a more distinctive three-dimensional pattern of anomalies, whereas a long strip of sea floor produces a two-dimensional pattern that is more ambiguous. While his supervisor Drummond Matthews, who had collected the data, was away, Vine conceived an explanation for the magnetic stripes ([7], p. 219).

Lawrence Morley in Canada was involved in aeromagnetic surveys over Canada, and was familiar with many aspects of geomagnetism. In his seafloor spreading paper, Dietz had commented on how the magnetic stripes off the western USA seemed to run under the continental slope, and had suggested that they were carried under the continent by subduction and destroyed by subsequent heating. It is well known that magnetisation does not survive if rocks are heated. Conversely, it is reacquired by magnetic materials upon cooling. Morley realised that the oceanic crust could be magnetised as it formed and cooled at a spreading ridge ([7], p. 217).

What has become known as the Vine–Matthews–Morley hypothesis combines the hypotheses of seafloor spreading and magnetic field reversals. The idea is that oceanic crust becomes magnetised as it forms at a spreading centre, and a strip of sea floor accumulates that records the current magnetic field direction (Figure 3.4(a)). If the magnetic field then reverses and the seafloor spreading continues, a new strip will form in the middle of the old strip (Figure 3.4(b)), the two parts of the old strip being carried away from the ridge crest on either side. Subsequent reversals would build up a pattern of normal and reverse strips, and the pattern would be symmetric about the ridge crest (Figure 3.4(c)).

Vine later commented that the hypothesis required three assumptions, each of which was, at the time, highly controversial: seafloor spreading, magnetic field reversals, and that the oceanic crust (the seismic 'second layer') was basalt and not

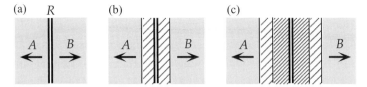

Figure 3.4. Illustration in map view of the way seafloor spreading and magnetic field reversals combine to yield strips of sea floor that are alternately normally and reversely magnetised. The resulting pattern is symmetric about the crest of the ridge if the spreading itself is symmetric (meaning that equal amounts of new sea floor are added to each plate).

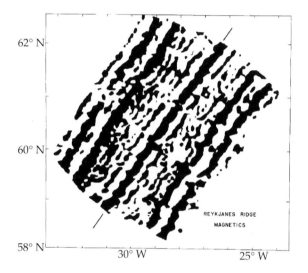

Figure 3.5. The pattern of magnetic anomalies across the Mid-Atlantic Ridge south of Iceland, where it is known as the Reykjanes Ridge. Black indicates a positive anomaly, inferred to be due to normally magnetised crust, and white indicates a negative anomaly, inferred to be due to reversely magnetised crust. The short lines mark the location of the ridge crest, along which there is a positive anomaly. Despite the irregularities, the pattern shows a striking symmetry about the ridge crest. From Heirtzler et al. [30]. Copyright by Elsevier Science. Reprinted with permission.

consolidated sediment ([7], p. 220). Morley submitted a paper about the beginning of 1963, which was rejected by two journals in succession, the second with unflattering comments. Vine and Matthews submitted a paper in about July 1963, which was published in September. Morley's story emerged later [7].

Subsequent exploration revealed extensive patterns of magnetic stripes on the sea floor, with an astonishing degree of symmetry about ridge crests (Figure 3.5), and which correlated with the field reversal chronology established on land. These magnetic stripes provided strong and startling evidence in favour of seafloor spreading. They also opened the prospect of assigning ages to vast areas of the sea floor on the basis of the reversal sequence, which was rapidly correlated from ocean to ocean.

We should reflect on the magnitude of that last paragraph. Assigning ages to rocks always has been and still is a central occupation of geologists. It is painstaking work, whether the method is correlation of fossils or measurement of radioactive decay. It took much of the twentieth century to develop the ability to get reliable ages accurate to within a few per cent or less for many kinds of rocks. As Menard remarked ([7], p. 212):

To general astonishment, magnetic reversals provide the long-sought global stratigraphic markers that are revolutionising most of geology. At sea, as though by a miracle, magnetic anomalies give the age of the sea floor without even collecting a sample of rock.

3.4 Evidence for motion – seismology

Seismology had already provided a key piece of evidence even before Wilson conceived of transform faults, as will be reiterated shortly. The Lamont (now Lamont–Doherty) Geological Observatory of Columbia University in New York state, directed by Maurice Ewing, had pioneered the exploration of the Atlantic sea floor, and then of other oceans. After Dietz' paper on seafloor spreading, Ewing turned much of the effort to testing the hypothesis. Part of this programme was to study the earthquakes in oceanic regions, and it was already known that these occur mainly on mid-ocean ridges. By 1963 there was a better distribution of modern seismographs around the world, including the newly deployed World-Wide Standardised Seismograph Network. This permitted earthquakes in remote regions to be located with an accuracy about ten times better than previously.

Lynn Sykes, working at Lamont, found that the earthquakes are located within a very narrow zone along the crests of mid-ocean ridges, and along the joining segments of fracture zones, where these were known or could be inferred [31, 32]. He made the explicit point that earthquakes on fracture zones occur predominantly on the segments that connect segments of ridge crest, and hardly at all on segments beyond ridge crests (Figure 3.6). This had been very puzzling when it was thought that fracture zones had offset ridges by motion along the length of the fracture zone. However, it was explicitly predicted by the transform fault concept, and was noted (barely) by Wilson as evidence in its favour (Figure 3.3).

When Sykes saw the evidence of his colleagues for symmetric magnetic anomalies, he was convinced of seafloor spreading, but realised that he could make another decisive test through seismology. The elastic waves emitted by an earthquake have

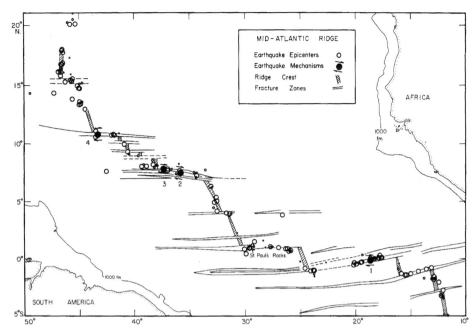

Figure 3.6. Earthquakes along the Mid-Atlantic Ridge. Open symbols show locations (epicentres) of earthquakes, and solid symbols with arrows show the sense of slip inferred from fault plane solutions. The fracture zones (oriented east–west) have earthquakes mainly between segments of the ridge crest, and only rarely on the extensions beyond ridge crests. These locations and the sense of slip on the active segments are consistent with Wilson's transform fault hypothesis. From Sykes [33]. Copyright by the American Geophysical Union.

a distinctive four-lobed pattern. In two opposite lobes, the waves that arrive first are compressional. In the intervening two lobes, the 'first arrivals' are dilatational. These waves spread through the Earth's interior in all directions. With a global distribution of seismographs, it is possible to sample these waves with sufficient density to reconstruct the orientation of the lobed 'radiation pattern' and the orientation of its two 'nodal planes'. One of these planes corresponds to the fault plane, and the other is perpendicular, though you can't tell which is which just from the seismic waves. It is also possible to infer directly the orientation of stresses at the earthquake source. The result of this determination was called a 'fault plane solution'.

Sykes knew that some previous fault plane solutions on ridges were suggestive, but that he could get much more reliable results from the new global seismographic network. This he did. He found that, for earthquakes located on segments of ridge crest, the solutions indicated normal faulting, consistent with the ridge crest being extensional. Earthquakes located on active segments of fracture zones had one

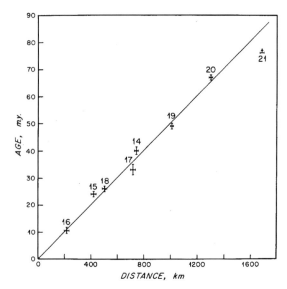

Figure 3.7. Ages of sediments immediately above the basaltic basement of the sea floor of the South Atlantic, plotted against distance from the crest of the Mid-Atlantic Ridge. The ages are inferred from micro-fossils. From Maxwell *et al.* [34]. Copyright American Association for the Advancement of Science. Reprinted with permission.

nodal plane approximately parallel to the fracture zone, consistent with strike-slip faulting (Figure 3.6). Most importantly, the sense of strike-slip motion was consistent with that predicted by the transform fault hypothesis, and opposite to that predicted by the simple transcurrent offset interpretation. This was another kind of observation strongly supportive of seafloor spreading.

3.5 Evidence for motion – sediments

Ewing, during the same period, had used seismic refraction to determine the thickness of sediments in the Atlantic. If seafloor spreading were occurring, the thickness of sediments should increase with distance from the ridge crest. The results were confusing, partly because of the rough seafloor topography of the Atlantic, and Ewing was reluctant to come out in support of seafloor spreading.

Later, a different approach became possible through a deep-sea drilling programme, which allowed the recovery of long sediment cores. An early cruise in the South Atlantic Ocean was aimed specifically at testing seafloor spreading. The results were spectacular. It was found that the age of the oldest sediment, just above the basaltic basement, determined from micro-fossils, increased in simple proportion to distance from the ridge crest, exactly as predicted by assuming seafloor spreading at a nearly constant rate (Figure 3.7). The results also provided an important calibration of the magnetic reversal chronology, which until then was well calibrated only for the first few million years.

Menard and others have remarked that most scientists are converted to a new idea by observations from within their own speciality. Thus palaeomagnetic polar wandering converted a small minority of geophysicists to continental drift. Later, the dramatic evidence of seafloor magnetic stripes, earthquake distributions and fault plane solutions converted a majority of geophysicists to seafloor spreading. To many of the more traditional geologists, however, such geophysical observations were still unfamiliar, and they were unsure how to regard them. However, fossil ages are a long-standing concept in geology, and something to which most geologists can readily relate. Thus, although the deep-sea sediment ages were not published until 1970, they were important for spreading the word to the great majority of geologists who work on continental geology.

This completes my short survey of some of the most direct and compelling evidence that led to the acceptance of plate tectonics by a majority of geologists. There is much other evidence and there were many more players, but a knowledge of these observations suffices to place plate tectonic motions on a firm empirical footing.

4

Solid, yielding mantle

> The mantle is solid and deforming, not molten. How do we know? Viscosity. Estimating the viscosity of the mantle. Temperature dependence of deformation of solids. Inevitable convection.

Prior to the twentieth century, geologists had deduced that the interior of the Earth had to be yielding, in order to accommodate the uplift of mountains. Some geologists assumed the interior was molten, but others acknowledged that a yielding solid would be sufficient. The clearest expression of the latter viewpoint is in the famous 1855 paper by Airy [35] in which he proposed an explanation for mountains being approximately in isostatic balance, namely that mountains have thickened crust beneath them. However, Airy also cited a key observation to support the idea of a deformable, solid interior [35]:

This fluidity may be very imperfect; it may be mere viscidity; *it may even be little more than that degree of yielding which (as is well known to miners) shows itself by changes in the floors of subterraneous chambers at a great depth* [emphasis added] when their width exceeds 20 or 30 feet [7 or 10 m]; and this degree of yielding may be sufficient for my present explanation.

There had been debate about whether the interior is presently fluid or whether it had only been fluid when the Earth was forming. It was Hall [36] in 1859 who established that continuing deformation was required, through his observations that sediments throughout thick sedimentary formations had all been deposited in shallow water. This required continuous subsidence as the sediments accumulated.

The development of seismology late in the nineteenth century led to the discovery that the mantle transmits two kinds of waves that were identified as compressional and shear. Liquids have no shear strength, and so cannot transmit shear waves. The mantle must be solid to transmit both compressional and shear waves. The apparent contradiction of a material being simultaneously solid and fluid is resolved by observing that the timescale of seismic waves is very short (tens of seconds) whereas the timescale of geological processes is very long (millions of years). Solids that yield over millions of years don't have time to yield in a few seconds. The term *liquid* refers to materials in the liquid state, like water and magma. The more general term *fluid* can serve to describe the slow yielding of a material in the solid state.

By early in the twentieth century a picture had emerged of a strong surface layer about 150 km thick that does not normally deform even on geological timescales, underlain by a more yielding interior. The strong layer had come to be called the *lithosphere*, and Barrell [8] in 1914 proposed the term *asthenosphere* for the weaker region below. I will not use the term 'asthenosphere' because its meaning became quite confused in the early days of plate tectonics. Some people thought there was a very weak, lubricating layer about 100 km thick immediately under the lithosphere, whereas others assumed convection was confined to the upper mantle, which by implication would be the asthenosphere. These days we have clear evidence for convection extending deep into the lower mantle, so the 'weak layer' includes much or all of the mantle. Therefore, I will speak only of the lithosphere and the convecting mantle. Even if there were a weaker layer under the lithosphere, it is hard to resolve and it would make little difference to the large-scale flows associated with plates.

The concept of the deformable mantle just described was still a qualitative concept. In order to quantify the 'mere viscosity' or 'degree of yielding', observations were needed of presently occurring deformations. These came from the 'rebound' of the Earth's surface following the removal of thick ice sheets at the end of the Ice Age, about 11 000 years ago. Before we consider those, we need a way to characterise a fluid's resistance to deformation.

4.1 Viscosity

In mechanical terms, a fluid is a material that can undergo an unlimited amount of deformation. A solid, on the other hand, may deform to a small extent, but it will break if you try to deform it too much. Another distinction is that many solids will deform only by a certain amount under the action of a particular force, and then return to their original shape if you stop applying that force. Such materials are called *elastic*. On the other hand, a fluid will keep deforming as long as a force is applied to it, and if the force is removed it will simply stop deforming, without returning to its original shape.

These distinctions are often very clear in our common experience, but in some circumstances they are not so clear. Thus, for example, some metals are elastic

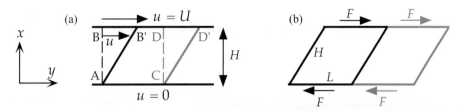

Figure 4.1. Shear flow in a layer of viscous fluid.

under the action of a small force, but yield and permanently deform if you apply a larger force. Malleable wire is a familiar example. A metal deforming permanently is behaving in part like a fluid. The tendency to behave more like a fluid is enhanced in many materials if we heat them, and metals again provide a familiar example. Even when a material is solid for all practical purposes, it may be undergoing very slow deformation, so we can consider it to be a fluid over hundreds or millions of years. For example, it is reported that the glass in some old cathedral windows has sagged detectably over hundreds of years.

A *viscous fluid* is a material whose rate of deformation is proportional to the applied force. We will look shortly at how we can quantify that statement. Viscous fluids are also known as *Newtonian* fluids. Strictly speaking, the term 'viscous' applies to materials in which the relationship between force and rate of deformation is *linear*, although the term is sometimes used more loosely. More general behaviour, such as that of malleable wire, is called variously *ductile*, *malleable* or *nonlinear*. Strictly speaking, ductile refers to materials with sufficient strength under tension that they can be stretched or drawn. Malleable would be a more appropriate term for many geological materials, but the term ductile is used fairly commonly.

In order to quantify our definition of a viscous fluid, we need ways to characterise a deformation and an applied force. The situation depicted in Figure 4.1 is about the simplest way to do this. Figure 4.1(a) depicts a layer of fluid between two horizontal plates. It may help to think of the fluid being 'stiff', 'thick' or 'gooey' like honey or treacle (molasses), rather than runny like water. Coordinates x and y are shown. Horizontal velocity is denoted as u and vertical velocity as v (though here it is zero everywhere). The top plate is moving to the right with velocity U, and the bottom plate is stationary. The fluid velocity increases steadily with height H in the layer, from 0 to U. Thus if we could quickly inject a line of dye along the line AB, it would at a later time become inclined like the line AB'. Similarly, the line CD will be carried into line CD'.

Fluids may deform by changing volume or by changing shape. They might also rotate, but this does not involve internal deformation. Although the mantle does

4.1 Viscosity

compress significantly under the extreme pressures at its base, we don't need to consider that here. The main effect of compression can be taken into account in a fairly simple way later if it is needed. So here we can consider our fluid to be incompressible and just focus on deformation.

The box defined by ABDC becomes *deformed* into the parallelogram AB'D'C. We can use this change in shape of the box to measure the deformation of the fluid. One way to measure the deformation of the box is with the ratio of lengths BB'/AB. If the time interval that has elapsed between when the dye is at AB and when it is at AB' is Δt, then the distance from B to B' will be $U\Delta t$. Then we might write

$$\text{shape change} = U \Delta t / H,$$

where H is the layer thickness. Then we would be using the tangent of the angle BAB' as our measure of deformation. For our viscous fluid we need the *rate* of change of the shape or the rate of deformation. This can be found by dividing by Δt, so

$$\text{rate of deformation} = U/H.$$

You can see that this quantity is a spatial *gradient of velocity* – in other words, it is the rate of change of (horizontal) velocity with (vertical) position. It may be a little confusing at first that the direction in which the velocity changes (vertical) is not the same as the direction of the velocity (horizontal), but if you look at Figure 4.1(a) it should make sense.

Velocity gradient turns out to be a generally useful quantity to measure rates of deformation. The technical terminology for a quantity that measures deformation is called *strain*. It follows that a quantity that measures *rate of deformation is called a strain rate*, so this is an alternative name for our velocity gradient. Here I will use the symbol s for strain rate. For consistency with the more general technical treatment that applies in two and three dimensions, which I will spare you here, I will include a factor of one-half in the definition of strain rate for Figure 4.1(a), so we get to the following definition of rate of deformation, or strain rate:

$$s = \frac{U}{2H}. \tag{4.1}$$

Now let's turn to the force causing the deformation. A force must be applied to the top plate in order to keep it moving. The moving plate then imparts a force into the adjacent fluid. The force imparted into the top of the deformed box is depicted in Figure 4.1(b) as F. The magnitude of this force depends on the length, L, of the box. For example, a second, adjacent box would also have a force F imparted into it, and the total force imparted into both boxes would have to be $2F$ in order to cause the same rate of strain. However, the deformation of each box is the same. Therefore what counts is the *force per unit area* that is applied to the fluid. We are familiar with pressure being a force per unit area, but pressure is a special case of the more general concept of *stress*, so I will use that term here. Thus the driving influence will be characterised as a stress.

As an aside, stress is more general than pressure because it allows that the force might be in different directions. For example, the force in Figure 4.1(b) is tangential to the top surface, and it generates a *shear stress*, but another force perpendicular to the surface might also act, generating the more familiar case of *pressure*. Also forces might act on the sides of the box as well as the top, and they need not be the same magnitude as the top forces. Each of these forces gives rise to a separate component of a *stress tensor*. However, we don't need to bother with tensors here.

We need to note at this point that Figure 4.1 is implicitly a cross-section through a structure that extends into the third dimension (out of the page). We can make this explicit by assuming that the box has a width W in the third dimension. Then the stress, τ, imparted to the top of the fluid is

$$\text{stress} = \frac{\text{force}}{\text{area}} = \tau = \frac{F}{LW}. \tag{4.2}$$

This quantity will serve as our measure of the applied force causing deformation.

We now have quantities that quantitatively characterise the rate of deformation and the driving force in Figure 4.1, so we are ready to look at the relationship between them. A viscous fluid is defined as one in which strain rate is proportional to stress. To be consistent with the more general technical development, I will again include a factor of 2 in the definition:

$$\tau = 2\mu s. \tag{4.3}$$

Equation (4.3) is called a *constitutive equation*; it describes the mechanical properties of a material. The constant of proportionality, μ, is called the *viscosity*. Viscosity is a material property that characterises a fluid's resistance to deformation. A fluid with a high viscosity requires a greater stress to produce a given rate of deformation. Since strain rate has the dimension 1/time and stress has the dimension force/area, or pressure, the units of viscosity are pascal seconds or Pa s (1 Pa = 1 N/m^2). Honey at room temperature has a viscosity in the range 10–100 Pa s. Water has a viscosity of about 0.001 Pa s. The mantle has a rather larger viscosity.

4.2 Viscosity of the mantle

We might expect that, if the mantle is made of solid but deforming rock, then its viscosity might be very large. But how large? A million times more viscous than honey? A billion times? Without some way of estimating it, we have trouble even guessing what order of magnitude it might be.

4.2 Viscosity of the mantle

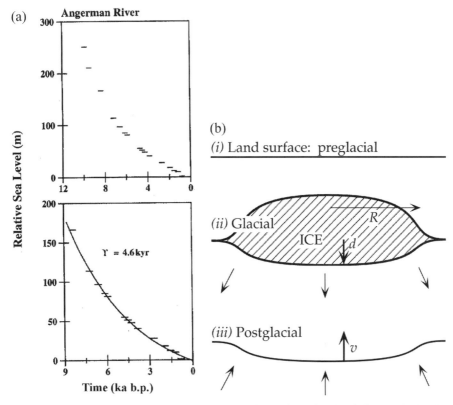

Figure 4.2. (a) Observations of the former height of sea level relative to the land surface at the Angerman River, Sweden. From Mitrovica [37]. Copyright by the American Geophysical Union. (b) Sketch of the sequence of deformations of the land surface (i) before, (ii) during, and (iii) after glaciation.

The land surfaces of Canada and of Scandinavia and Finland (referred to as Fennoscandia) have been inferred to be rising at rates of millimetres per year relative to sea level. The main observation on which this inference is based is a series of former wave-cut beach levels raised above present sea level. These have been dated in a number of places to provide a record, which is usually presented as relative sea level versus age. An example is shown in Figure 4.2(a).

The interpretation of these observations of recent and continuing uplift is that the land in these areas is rising because the weight of thick ice sheets that were present in the last Ice Age has been removed. The implication is that the ice depressed the land surface as it accumulated, and that the land surface rises or 'rebounds' after the ice has melted. The fact that the rebound is protracted rather than instantaneous indicates that the mantle under these areas is behaving as a highly viscous fluid rather than just as an elastic solid.

The ice sheets melted rather rapidly around 11 000 years ago. Since that time there has been a clear slowing of the rate of uplift, as can be seen in Figure 4.2(a).

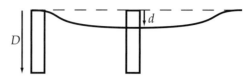

Figure 4.3. Columns with a cross-sectional area of 1 m².

In fact, the lower panel shows some of the data with an exponential curve fitted through them. The exponential has a time constant of 4.6 kyr (kiloyears; this is the time for the uplift to decrease by a factor e).

The inferred sequence of events is sketched in Figure 4.2(b). An initial flat reference surface (i) is depressed a distance d by the weight of glacial ice during the Ice Age (ii). (The ice load peaked about 18 kyr before present and ended about 10 kyr.) After melting removed the ice load, the reference surface started rising back towards its isostatically balanced level (iii), and that rising continues at present with velocity v. In order for the surface to move vertically, the underlying mantle must flow away, at first, and then towards the depression, as indicated by the arrows.

Contrary to the common image of physics as being an 'exact' science (there is no such thing), or of using sophisticated mathematics to reach conclusions of high precision, a lot of good physics can be done with very rough and simple calculations, especially for the Earth, where our information often has large uncertainties. The post-glacial rebound problem has been analysed using the partial differential equations governing viscous fluid flow, and taking account of many details of the situation, and from such analyses the viscosity of the mantle under Fennoscandia has been estimated. On the other hand, a fair estimate of the viscosity of the mantle can be obtained through the following rough approximation, using the definitions of stress, strain and viscosity given in the previous section.

The depression was caused by the weight of ice sheets 2–3 km thick. If the ice sheet were in place for long enough, an isostatic force balance would have been established between the ice pushing down and the mantle 'pushing' up. When the ice melted, an imbalance of forces was created, because the mantle was still pushing up. A more rigorous way to say it is that there was a pressure deficit in the mantle under the depression, because the depression was filled by water or air rather than by heavier rock. We can calculate that pressure deficit by considering the weight of two vertical columns of equal depth and with a horizontal cross-section of 1 m², as is illustrated in Figure 4.3.

The left-hand column is outside the depression and it contains all rock. Its volume is $D \times 1 \times 1 = D\,\mathrm{m}^3$, so its mass is $\rho_m D$ and its weight is $g\rho_m D$, where ρ_m is the density of the mantle (it turns out that the effect of the crust cancels

4.2 Viscosity of the mantle

out of the two columns and only the mantle density counts). Thus the force per unit area, or pressure P_l, at the base of the left-hand column is $g\rho_m D$, the weight exerted on the base of area 1 m². Suppose the depression is filled with water, of density ρ_w. Then the right-hand column, which is in the middle of the depression, has water to a depth d and rock for the rest of its depth, which is $D - d$. Thus its mass is $\rho_w d + \rho_m (D - d)$ and its weight is g times that mass. Now we can subtract the pressure exerted by the right-hand column from the pressure exerted by the left-hand column to obtain the pressure deficit, ΔP, under the depression:

$$\Delta P = P_l - P_r = g\rho_m D - g[\rho_w d + \rho_m (D - d)]$$
$$= g(\rho_m - \rho_w)d.$$

We can think of this as g times the mass deficit created by replacing rock with water in the depression. Now pressure is one form of stress, so we can write

$$\tau_d = g(\rho_m - \rho_w)d. \qquad (4.4)$$

This is, to a rough approximation, the stress that drives the flow in the mantle that must occur as the floor of the depression rises to the level it had before the Ice Age. That flow is depicted by the arrows in Figure 4.2(b).

This driving stress or pressure acts over the area of the depression, which is about πR^2, assuming the depression is roughly circular with a radius of R. If pressure is force per unit area, then pressure times area yields a total force, the driving force

$$F_d = \pi R^2 g(\rho_m - \rho_w)d. \qquad (4.5)$$

The driving force is resisted by a force arising from viscous stresses in the mantle. According to Eq. (4.3), viscous stress is proportional to strain rate, and according to Eq. (4.1) strain rate is proportional to velocity gradient. If we can identify a typical velocity gradient within the mantle, then we can estimate the viscous resisting stress. If the rate of uplift of the floor of the depression is v, then it is plausible that the upward velocity of the mantle immediately under the depression is close to v. It is also plausible that as you go away from the depression the uplift velocity declines, so that far from the depression the velocity is close to zero. How far do you have to go for the velocity to drop substantially? That will depend on the width of the depression. The Fennoscandian depression has a radius of about 1000 km, so if you go 1000 km away the velocity should have decreased significantly. On the other hand, the North American depression has a radius of about 2500 km, so its effect in the mantle should extend further, vertically and horizontally, and you would have to go 2500 km for the velocity to drop significantly. This suggests that a representative velocity gradient within the mantle flow is v/R, in the sense that the velocity changes by a significant fraction of v over a distance R. It may be that you have to go a distance $2R$ for the velocity to drop by, say, 80%, but let's not worry about factors of 2 for the moment.

If we have a representative velocity gradient of v/R, then from Eq. (4.1) a representative strain rate is $s = v/2R$. Then from Eq. (4.3) a representative viscous stress is

$$\tau_r = 2\mu s = 2\mu v/2R = \mu v/R. \tag{4.6}$$

This is the stress within the mantle that resists the driving stress due to the pressure deficit, as given by Eq. (4.4). It also acts over an area that is comparable to the area of the depression, so we can say that the resisting force is roughly

$$F_r = \pi R^2 \mu v / R. \tag{4.7}$$

The driving and resisting forces must balance. This follows from Newton's second law of motion, which says that force = mass × acceleration. If acceleration is zero, then (net) force must be zero. Strictly speaking, the uplift velocity is changing, according to Figure 4.2(a), so the acceleration is not exactly zero. Yet the velocity is very small, and it changes only over thousands of years, so the acceleration is tiny. Let's put numbers to those statements. From Figure 4.2(a), about 100 m of uplift has occurred over about 10 000 years, so the average uplift rate is about 1 cm/yr. A year is about 3.12×10^7 seconds (go on, work it out for yourself), so the velocity is about 3×10^{-10} m/s. This velocity changes over about 3000 years, which is about 10^{11} s, so the acceleration is about 3×10^{-21} m/s^2, a very small quantity. It is characteristic of mantle flow that accelerations are totally negligible. It follows that everywhere in the mantle forces and stresses balance, to a very good approximation.

So the driving force should equal the resisting force, or

$$F_d - F_r = 0,$$

so

$$\pi R^2 g(\rho_m - \rho_w)d - \pi R^2 \mu v/R = 0.$$

We can rearrange this to give

$$\mu = g(\rho_m - \rho_w)dR/v. \tag{4.8}$$

Using the values already mentioned as well as $\rho_w = 1000$ kg/m^3 and taking d to be 100 m yields $\mu = 8 \times 10^{21}$ Pa s. A more rigorous solution of Eq. (4.8) is given in Appendix A (Section A.2), and it yields $\mu = 3.3 \times 10^{21}$ Pa s. The original analysis by Haskell [38] in 1937 for the Fennoscandian region yielded 10^{21} Pa s. A recent analysis by Mitrovica [37] in 1996 confirmed Haskell's value, though it also demonstrated that the viscosity is lower in the upper few hundred kilometres of the mantle and larger at greater depths.

At the beginning of this section I posed the question of how large the mantle viscosity is. These results show that it is very much larger even than a billion times the viscosity of honey – something like a thousand billion billion times. The earlier guess was wrong by 12 orders of magnitude! Even though the estimate made here is very rough, it has given the right order of magnitude, though it differs by a factor of 8 from Haskell's value. Thus our rough estimate is very valuable, because it has replaced serious ignorance with some useful knowledge. It is important, of course, to bear in mind that the estimate is not expected to be very accurate, but it is reasonable to believe (even without comparisons from other studies) that it is accurate to better than one order of magnitude. There is another virtue of the very simple approach used here, which is that it allows the physics to remain clearly in view, without being obscured by mathematics.

4.3 Dependence of viscosity on temperature

You probably know that cool honey is more viscous that warm honey. Honey that has been kept in a refrigerator is much more viscous than honey at room temperature on a hot day. Even a temperature change from 20 °C to 30 °C might change the viscosity by an order of magnitude or so. It turns out that rocks at the high pressures and high temperatures of the mantle behave similarly. An increase of temperature from 1300 °C to 1400 °C can reduce the viscosity by about a factor of 5.

The reason for this strong sensitivity to temperature is that rock deformation occurs by the movement of defects within the crystalline structures of the rock's minerals. Atoms are constantly in motion, jiggling around their mean position in the crystal, and the amount of jiggling increases with temperature. The movement of defects depends on atoms occasionally jiggling into a neighbouring position, and the probability of such a jump increases rapidly as the amplitude of the atom's jiggling increases. This is a well-studied process, called thermal activation, and thermally activated processes have a typical form of dependence on temperature. This form, calibrated by experiments, can be used to deduce the following dependence of viscosity on temperature T [39]:

$$\mu = \mu_r \exp\left[\frac{E^*}{R}\left(\frac{1}{T} - \frac{1}{T_r}\right)\right], \tag{4.9}$$

where μ_r is the viscosity at a reference temperature T_r, R is the universal gas constant and E^* is called the activation energy.

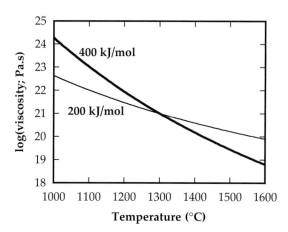

Figure 4.4. Dependence of viscosity on temperature. Each curve is labelled with its activation energy.

Examples of the dependence of viscosity on temperature for two different activation energies are shown in Figure 4.4. Activation energies are not easy to determine, and there is some debate about which value applies to the upper mantle. The value also depends on whether the mantle is really a linear viscous fluid or has a more nonlinear dependence on stress [39]. Pressure tends to increase the activation energy (strictly speaking, it increases the activation enthalpy), so the sensitivity of viscosity to temperature is likely to be even stronger in the deep mantle.

4.4 Inevitable convection

This strong dependence of mantle viscosity on temperature plays a very important role in mantle convection. It is the reason the lithosphere is much stronger than the underlying mantle. It also controls the form of mantle plumes, as we will see in Chapter 7. Also, Tozer [40] argued in 1965, at a time when convection throughout the mantle was still controversial, that the temperature dependence of mantle viscosity made convection virtually inevitable in an Earth-sized planet.

Tozer's argument was that the planet would presumably be heated to some degree by the release of gravitational energy as it accreted from fragments orbiting the Sun. Radioactivity would then slowly heat it more. The conductivity of rocks is low, and conduction would only cool the planet to a depth of around 500 km over the age of the Earth. Sooner or later the deeper interior would become hot enough that the mantle viscosity was reduced to a value that permitted convection. Thereafter the mantle temperature and viscosity would self-regulate to remove whatever heat

4.4 Inevitable convection

was being generated within it. Thus if the mantle should be overcooled, it would become more viscous, its convection would slow, its heat loss would slow and it would begin to warm up again. Conversely if the mantle became overheated, its viscosity would drop, convection and heat loss would speed up and it would cool again. This was a very sound argument that was borne out by later calculations of the thermal evolution of the mantle, which we will see in Chapter 9.

5

Convection

> Thermal convection is driven by boundary layers. Buoyancy. How to calculate plate velocities – simple mechanical version. Interpretation of an oceanic plate as a thermal boundary layer. Estimation of plate thickness from thermal conduction. Plate velocities – thermo-viscous version. What is a Rayleigh number? Other useful numbers.

Convection is not familiar to many geologists, so before we get to mantle convection, which has some unusual features, we need to get to know convection in general. If we start from the right place, then convection becomes understandable in straightforward terms. Instead of it being something vague down there that makes things go around, we can know when to expect convection and what will control it.

The key to convection is that it is driven by boundary layers. I will focus for the moment on thermal convection (there is also compositional convection), so thermal convection is driven by thermal boundary layers. If we can identify a thermal boundary layer, and say something about how it behaves, we are well on the way to understanding the convection that it drives.

In the mantle, the relevant thermal boundary layers form at horizontal boundaries. Later we will look at why that is. For the moment, we can use Figure 5.1 as the simplest realisation of what I have just said. There is a hot thermal boundary layer along the bottom boundary of the fluid layer. It is common in textbooks to depict thermal convection with two thermal boundary layers, a hot one at the bottom and a cold one at the top. However, this is only one of a range of possibilities, and it is not exactly the situation in the mantle. Therefore I have kept the situation as simple as possible in Figure 5.1 to emphasise that there may be one, or the other, or both thermal boundary layers.

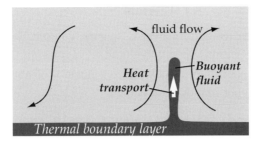

Figure 5.1. Essential elements of convection in a fluid layer.

The other fundamental thing to say about convection is that it is driven by buoyancy. Buoyancy arises from the action of gravity on density differences. The density differences can be due to differences in composition or differences in temperature – this gives us the distinction between compositional convection and thermal convection. Differences in temperature cause differences in density because of the phenomenon of thermal expansion. Thus warmer fluid expands, its density is lower, its weight (per unit volume) is lower, so it rises relative to other fluid. Conversely, cooler fluid contracts, its density is higher, it becomes heavier and sinks. Thus we get to the familiar phenomenon that hot air rises.

If we put these two things together, the fluid in a thermal boundary layer is buoyant (positively or negatively), its buoyancy can cause it to rise (or sink), and the flow driven by that rising or sinking fluid is convection. A rising column of buoyant fluid is depicted in Figure 5.1, and the flow it induces is shown by the thin arrows. Because some fluid is rising, other fluid must sink to compensate. However, this return flow does not have to occur in simple closed circulation patterns, if the flow is time-dependent or three-dimensional. Because warmer, rising fluid carries extra heat within it, it transports that heat upwards through the fluid. Thus thermal convection is also a heat transport mechanism. That is the essence of thermal convection. This overview can be a guide as we go through the specifics of each part of the process, and then put them together into a description of thermal convection.

There are several material properties that are important in convection. We have just encountered thermal expansion, which depends on the particular material involved, which is why it is called a material property. We have also encountered density, another material property. Viscosity was covered in the previous chapter. It is a measure of how much the fluid resists deformation, and we need that to figure out how fast the fluid will flow under the action of a given buoyancy. Later we will need the thermal conductivity, which controls the thickness of thermal boundary layers. Specific heat will also crop up, but we'll worry about that when we get to it.

5.1 Thermal expansion

We will start with density, and its inverse, specific volume. Density, ρ, is the mass of a unit volume of material. Specific volume, V, is the volume occupied by a unit mass of material. Upper-mantle rocks have a density of about $3300\,\text{kg/m}^3$. The specific volume of upper-mantle rock is then

$$V = 1/\rho = 1/3300\,\text{m}^3/\text{kg} = 3 \times 10^{-4}\,\text{m}^3/\text{kg}. \tag{5.1}$$

We more commonly work with density, but volume may be easier to visualise for the moment.

Thermal expansion occurs when material warms up. Suppose it starts with a volume V_0 and a temperature T_0, then it warms up to temperature T and attains a volume V. The way thermal expansion works is that the change in volume is proportional to the initial volume and to the change in temperature. We can write this as

$$V - V_0 = \alpha V_0 (T - T_0). \tag{5.2a}$$

We can also write it as

$$\frac{\Delta V}{V_0} = \alpha \Delta T, \tag{5.2b}$$

where $\Delta V = V - V_0$ and $\Delta T = T - T_0$. Thus ΔV is the change in volume, and $\Delta V/V_0$ is the *fractional* change in volume. The quantity α in Eqs (5.2) is called the *coefficient of thermal expansion*. It is different for every material and thus it is called a material property. So, from Eq. (5.2b), we can say that the fractional change in volume is proportional to the change in temperature, and the constant of proportionality is the coefficient of thermal expansion.

Thermal expansion can also be expressed in terms of changes in density. Thus, from Eq. (5.2a)

$$\frac{1}{\rho} - \frac{1}{\rho_0} = \frac{\rho_0 - \rho}{\rho \rho_0} = \frac{\alpha}{\rho_0} \Delta T,$$

which can be rearranged into

$$\frac{\rho_0 - \rho}{\rho} = \frac{\Delta \rho}{\rho} = \alpha \Delta T.$$

This is close to the form of Eq. (5.2b), but there is a minus sign and, more significantly, the denominator is the *final* density, rather than the initial density. The minus sign arises because we choose to define the change in density as the final density minus the initial density. The second difference, the final density in the denominator, would complicate the algebra when using the relationship. Fortunately, the density changes in the mantle context (and in most convection) are quite small, of the order of 1%. Thus we would not make much of an error if we used the initial density instead of the final density in the denominator. Since it simplifies the later mathematics quite a lot, we normally do that, and get

$$\frac{\Delta\rho}{\rho_0} = -\alpha\Delta T. \qquad (5.3)$$

Now (5.3) is very similar in form to (5.2). The minus sign means that an *increase* in temperature causes a *decrease* in density. Just to be clear, the approximation we made, replacing the final density with the initial density in the denominator, causes a 1% error in $\Delta\rho$, not a 1% error in ρ. If the error were 1% of ρ, that would imply a 100% error in $\Delta\rho$, which would be serious, but that is not what we did.

In the upper mantle, α is about $3 \times 10^{-5}\,°C^{-1}$. Notice that the units are *per degree Celsius*. When that is multiplied by a temperature difference in Eq. (5.3) the result is dimensionless. This matches the left-hand side, which is a ratio of densities and therefore also dimensionless.

The temperature in the upper mantle is about 1300 °C. Because the temperature at the Earth's surface is around 0 °C, this means that a mantle rock taken from the surface to the upper mantle would undergo a fractional change in density, using Eq. (5.3), of

$$-\alpha\Delta T = -3 \times 10^{-5} \times 1300 = -3.9 \times 10^{-2}$$

or -3.9%. This is the largest thermal density difference we will meet in mantle convection. The absolute change in density is $-0.039 \times 3300 = -130\,\text{kg/m}^3$.

Notice that I have only been giving results to two significant figures. This is sufficient, since the thermal expansion is not known to any greater accuracy. This is another reason why the approximation made in getting to Eq. (5.3) is not a serious problem.

5.2 Buoyancy

A rock is denser than water, and if you release the rock in water it will sink. However, the downward force on the rock due to gravity is less than if the rock were in air. You can compare the vertical pressure gradients inside and outside the rock and reach this conclusion. However, the simple result is that gravity acts, in effect, only on the *difference* between the density of the rock and the density of water. It is an extension of Archimedes' principle: the net force is the weight of the rock minus the weight of the water displaced by the rock.

Ice is less dense than water, and if you release a piece of ice under water it will rise. In this case we say the ice is buoyant. In fluid dynamics, the technical term buoyancy, B, refers to the *force* due to the action of gravity on the *density difference* between volumes within the fluid. Because buoyant things go up, we define buoyancy to be positive when the force is upwards. Thus the buoyancy of a piece of ice of volume V in water is

$$B = -g(\rho_i - \rho_w)V, \tag{5.4}$$

where g is the acceleration due to gravity, and ρ_i and ρ_w are the densities of ice and water, respectively (V is now the total volume of the ice, not the specific volume used in the previous section).

A volume of warmer fluid will be less dense than its cooler surroundings and therefore buoyant. In this case we can write, using Eq. (5.3),

$$B = -g\Delta\rho V = g(\rho_0 \alpha \Delta T)V. \tag{5.5}$$

For example, suppose we approximate the buoyant column in Figure 5.1 as a cylinder of height $h = 2000$ km and diameter $r = 50$ km. Suppose it is 300 °C hotter than the surrounding fluid. Then its buoyancy is

$$B = g(\rho_0 \alpha \Delta T)(\pi r^2 h) = 4.6 \times 10^{17} \text{ N}.$$

Here I have used $g = 9.8$ m/s^2 and other values as previously specified. The units of force are newtons, N. This value of B is a large number, but it does not have much meaning until it is related to other things.

5.3 Plate velocity – simple mechanical version

Figure 5.2(a) is a sketch of a system of tectonic plates, in cross-section, with the middle plate subducting. Because the plate sinks, it displaces mantle material and thus induces flow in the mantle. The plate is taken to have a thickness d and to sink with velocity v. The mantle has an internal temperature T, a viscosity μ and a depth D. The surface temperature is $T = 0$. If the plate is sinking because it is cold and heavy, then this is an example of convection as it was defined above. In this case there is a cold thermal boundary layer at the top of the fluid, and the active component is a negatively buoyant sheet of cold material sinking away from the thermal boundary layer, whereas in Figure 5.1 it is the buoyant rising column that is the active component.

Figure 5.2(b) is a simplification of the situation in Figure 5.2(a). It may seem crude, but we can use the same approach here as we used in estimating the viscosity of the mantle in Chapter 4. I want to pose the question: 'How fast will the subducting plate move?' If you knew nothing about plate tectonics, you would have little idea of the answer. It might be kilometres per year, or millimetres per million years, or almost anything. By using the very simplified situation in Figure 5.2(b) we can make an estimate that will give us some idea of the answer. If we choose representative

5.3 Plate velocity – simple mechanical version

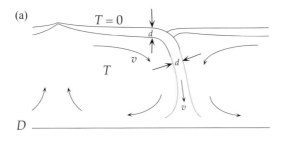

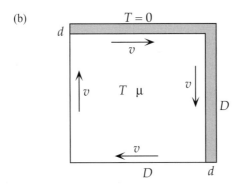

Figure 5.2. (a) Sketch of a subducting plate and the mantle flow it induces. (b) Simplification of the situation in (a).

values carefully, this estimate is likely to be within an order of magnitude or less of the accurate answer.

So, in Figure 5.2(b), the cooler, denser fluid in the sheet on the right-hand side of the box will sink. This will displace fluid within the box and thus induce a circulation in the box. It will also drag the surface plate along behind it. Referring back to Figure 5.2(a), the idea is that, as the sheet reaches the bottom, it spreads across the bottom, and, as the surface plate moves to the right, material is added to it at the spreading centre on the left. In this way the surface plate and its sinking component are continuously renewed and the picture remains basically unchanged.

I have stated that the vertical sheet will sink because it is heavy – in other words, it has negative buoyancy. As portrayed in Figure 5.2(b), there is nothing else that might drive flow in the fluid. Therefore the sinking sheet is the active component, and it will exert a force on the adjacent fluid. The fluid will resist motion because it is viscous. Therefore, there will be a resisting force exerted on the sinking sheet. Because mantle motion is so slow, accelerations are negligible, as discussed in Chapter 4. If the acceleration is zero, Newton's second law says that the net force on the sinking sheet must be zero. In other words, the resisting force, F_R, must balance the driving buoyancy force, B:

$$B + F_R = 0. \tag{5.6}$$

To calculate the driving force, the buoyancy, note that the volume of the sinking sheet is $V = D \times d \times 1 = Dd$. Because the sketch is a cross-section, the sheet implicitly extends in the third dimension. Therefore, let us calculate the forces on a unit length in the third dimension; that is where the factor 1 came from in the volume. The buoyancy will depend on the average reduction in temperature compared with the mantle. The temperature in the descending sheet is inherited from the plate at the surface. The temperature in the surface plate is zero at the Earth's surface and T at its base. Therefore, let us use the simple approximation that the average temperature in the plate is $(T+0)/2 = T/2$. Then the reduction in temperature $\Delta T = (T/2 - T) = -T/2$. So, from Eq. (5.5), the buoyancy is

$$B = -g\rho_0 \alpha T D d/2. \tag{5.7}$$

Therefore the buoyancy is negative – in other words, it is a downward force.

You may be worrying that the sheet will warm up as it descends, so we shouldn't use a constant temperature of $T/2$ throughout its depth. However, there is a good reason to do this. It is because the sheet will warm up by absorbing heat from the surrounding hot mantle, so the adjacent mantle will be cooled and itself become negatively buoyant. The effect is that the amount of heat is not changed, it is just redistributed horizontally, so we do not make a large error by treating it as though it is still in the descending sheet. Actually I could more accurately have said the *deficit* in heat is not changed.

Now let's consider the resisting force. This is due to the viscosity of the mantle, so it is proportional to the relevant velocity gradient, according to the discussion in Chapter 4. In Figure 5.2(b), the fluid on the right-hand side is descending at velocity v, whereas on the left-hand side it is rising at velocity v. Therefore the vertical velocity changes by the amount $2v$ over a horizontal distance D (assuming the box is square, for simplicity). Thus there is a velocity gradient of $2v/D$. This velocity gradient is an expression of the fact that the fluid is being sheared, i.e. deformed, in the sense of the vertical arrows in Figure 5.2(b). This deformation will induce a stress τ in the fluid, which, from Eqs (4.1) and (4.3) is

$$\tau = 2\mu v/D.$$

This stress acts on the descending sheet and thus generates a resisting force. The force is the stress times the area over which it acts. For a unit length of the sheet in the third dimension, that area is $D \times 1 = D$. Thus the resisting force acting on the surface of the descending sheet is

$$F_R = D2\mu v/D = 2\mu v.$$

This force acts upwards and is therefore positive. Actually Figure 5.2(a) suggests that we might count resistance in other parts of the system as well. For example, the fluid on the other side of the descending sheet will exert a similar resistance, and the fluid will also resist the motion of the surface plate, because of the shearing depicted by the horizontal arrows. In our idealised square box, these three forces are the same. Thus we might estimate the total resisting forces as the remarkably simple expression

$$F_R = 6\mu v. \tag{5.8}$$

We have now evaluated both terms in the force balance equation (5.6), so we get

$$-g\rho_0 \alpha T D d/2 + 6\mu v = 0,$$

which rearranges into

$$v = g\rho_0 \alpha T D d/12\mu. \tag{5.9}$$

Here we have a formula for the velocity of the sinking sheet, which in the square-box approximation is the same as the velocity of the surface plate. We can evaluate it using values already used, plus the observation (from seismology) that plates are about 100 km thick, so we can use $d = 100$ km. We also need the depth of the mantle, $D = 3000$ km, and we will use a viscosity that takes account of the fact that the viscosity is greater in the lower mantle, i.e. 10^{22} Pa s [41]. This gives a velocity of 3×10^{-9} m/s, which is about 10 cm/yr.

Observed plate velocities range from about 2 to 12 cm/yr, with an average of about 5 cm/yr. Our estimate is within a factor of 2 of the observed average. Thus our initial deep ignorance (Are plate velocities millimetres per million years, or kilometres per year, or something else even?) has been replaced with an estimate that is better than an order of magnitude in accuracy, even though our estimate is based on a very crude analysis. Again, this shows the value of even a very simple physical analysis, provided reasonably representative values are used.

5.4 Heat conduction

We can actually do better than the simple theory of the previous section. We evaluated the thickness of the plate, d, by appealing to observations. But in convection the thickness of the thermal boundary layer is determined internally, as part of the process of convection, so the theory ought to determine d without any extra input. We will extend our convection theory to do this in the next section, after we have looked at heat conduction, which is also known as *thermal diffusion*. There are several ideas to walk through, but bear with it because we get to a simple relationship that is quite powerful because it can be applied to many situations to deduce important things like timescales and depth scales.

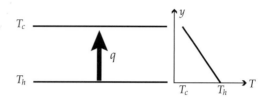

Figure 5.3. Conducted heat flow, q, is proportional to the temperature gradient.

Heat flows from hotter things to colder things. Experiments have established that the rate of heat flow is proportional to the local temperature gradient. For example, in Figure 5.3, the bottom of a layer of material is at a high temperature, T_h, and the top is at a cool temperature, T_c. Suppose the temperature varies linearly through the layer, as depicted in the graph of T versus height y on the right. Suppose also the layer thickness is h. Then the rate at which heat flows, q, will be

$$q = -K(T_c - T_h)/h. \tag{5.10}$$

I have written the temperature difference as the temperature at the larger value of y, minus the temperature at the smaller value of y, so that a positive temperature gradient will correspond to a positive slope in the graph. The slope is actually negative, yet the heat flows in the positive direction. Therefore the minus sign is required in front, to ensure that heat flows from hot to cold. The constant of proportionality, K, is the *thermal conductivity*, another material property. Equation (5.10) expresses the earlier statement that the rate of heat flow is proportional to the local temperature gradient. It is sometimes known as Fourier's law. In Eq. (5.10), q is actually a heat *flux*, the rate at which heat flows through a unit area of the surface of the material. Heat is a form of energy, so its units are joules, J. A rate of heat flow is measured in joules per second, or watts, W. Thus the units of q are W/m^2.

In order to deal with thermal boundary layers, we need to be able to consider how temperatures change with time (in Figure 5.3 the temperature is steady, even though heat is flowing through the layer). To do this, consider a hot body placed against a cold boundary, as depicted in Figure 5.4(a). Initially the body has a uniform temperature T_m and extends downwards from depth $z = 0$. The temperature at the surface, $z = 0$, is maintained at $T = 0$. You can think of this as the temperature profile through new oceanic crust and mantle that has just formed at a mid-ocean ridge spreading centre: hot mantle wells up and comes into contact with cold sea water. The temperature profile after contact is shown in Figure 5.4(b) by the short-dashed line labelled $t = 0$.

The question we now address is how the temperature profile will change with time. Heat will flow from the hot body through the cold boundary, so at a later time t_1 the temperature profile might be something like the solid line shown in

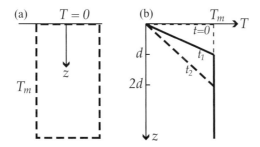

Figure 5.4. (a) A body of hot material placed against a cold boundary. (b) Sketch of the temperature variation with depth, z, in crude approximation, as it develops in time. The profile is shown at three times, as marked.

Figure 5.4(b). The logic of this is that heat will initially flow down the very steep gradient at the surface, but heat will not flow at depth because the temperature is still uniform down there: zero gradient gives zero heat flow, according to Eq. (5.10). Thus heat will be removed from near the surface, but the surface will still be at $T = 0$, and deeper down the temperature will still be T_m, so the profile must connect these two temperatures. We are simplifying by assuming that the profile is simply linear in this shallow range. A more realistic profile will be shown later. We can repeat this logic: more heat will flow down the new, gentler gradient and the cooling region will extend downwards, but at some greater depth the temperature will still be uniform. Thus the temperature at a later time t_2 will look something like the long-dashed profile shown in Figure 5.4(b).

The region where the temperature changes from the boundary value to the interior value is called a thermal boundary layer. Thus it is a transitional region, and typically the temperature gradient is relatively steep within it. We want to know how the thermal boundary layer will develop with time.

Now, to get at the rate at which the temperature profile changes, we can ask how long it has taken to go from the initial profile to that at time t_1. We can get a timescale by estimating a rate at which heat is being lost and combining that with the *amount* of heat lost. The rate at which heat is being lost at time t_1 is

$$q_1 = K T_m / d. \tag{5.11}$$

The rate of heat loss at earlier times will be greater than this, because the gradient is steeper, so this is a *lower bound*. If we also know how much heat has been lost, we can get an *upper bound* on the time it has taken to lose the heat.

The heat content of a body is determined by its temperature, its mass and a material property called the *specific heat*, C_P. This is the amount of heat it takes to raise the temperature of a unit mass of the material by 1 °C. For mantle rocks this is around 1000 J/kg °C. (The subscript P specifies that this is the specific heat at constant pressure, which allows the material some thermal expansion as it is warmed by one degree. Thermodynamicists also define the specific heat at constant volume, C_V.) The temperature of the material in the thermal boundary layer has decreased by an average of $T_m/2$. Therefore, the change in heat content per unit mass is $C_P T_m/2$. Density is mass per unit volume, so the change in heat content per unit volume is $\rho C_P T_m/2$. Now the heat loss in Eq. (5.11) is a heat flux, i.e. the rate of heat loss *per unit area* of the top surface (the sea floor, if this is the oceanic lithosphere). Therefore we need the change in heat content per unit area. This will be the change in heat content of a column of rock of depth d and unit cross-sectional area, which has a volume of $d \times 1 \times 1$. Thus we finally get the amount of heat lost per unit area of the surface as

$$\Delta H = \rho C_P T_m d/2. \tag{5.12}$$

If this amount of heat were lost at the rate given by Eq. (5.11), it would take a time

$$t_s = \Delta H/q_1 = \rho C_P T_m d^2 / 2 K T_m = \rho C_P d^2 / 2K, \tag{5.13a}$$

$$t_s = d^2/2\kappa. \tag{5.13b}$$

There are several things to say about this series of expressions. Temperature cancels out: the time does not depend on the temperature difference, only on the depth and the material properties. The material properties have been combined in the last expression into a quantity called the *thermal diffusivity*:

$$\kappa = K/\rho C_P. \tag{5.14}$$

A typical rock conductivity is about 3 W/m °C. With a density of around 3000 kg/m^3 and a specific heat of 1000 J/kg °C, we get $\kappa = 10^{-6}$ m^2/s. You can work through the dimensions of these quantities to verify that κ has dimensions of length^2/time, which it must have for the dimensions in Eq. (5.13b) to balance.

How long would it take to cool the body to a depth of 10 m? Without prior knowledge, you would not have much idea. It might be minutes or months. If we use the value of κ already given, the answer is 5×10^7 s, which you might recall is over a year. Well, this is an upper bound, but it indicates that it takes quite a long time to cool rocks to a depth of 10 m. The foundations of houses in cold climates are dug about 2 m into the ground so they extend below the level that freezes in winter. This is why basements are common in cold climates.

How long would it take to cool the body to a depth of 20 m? Because of the d^2 in Eq. (5.13b), it takes four times as long: 2×10^8 s or over 6 years. Why does doubling the depth cause the time to increase by a factor of 4? If you go through the earlier logic for the layer of depth $2d$ in Figure 5.4, you find that the rate of heat

5.4 Heat conduction

Figure 5.5. Sketch of the actual temperature profile of a cooling body.

loss is only half that for the thinner layer, because the temperature gradient is only half, so that would double the time. However, there is also twice as much heat to be removed, because a larger volume of material has to cool. That also contributes a factor of 2. Thus there is twice as much heat to be lost, but it is lost at only half the rate, so it takes four times as long.

The time t_s in Eq. (5.13) is an upper bound, but it still gives us an understanding of how the cooling time *scales*, i.e. how the time changes if the depth is changed, or if the material properties are changed. This is because the actual cooling time is a constant fraction of the upper bound t_s. This makes some sense if you note that the logic for the layer of thickness $2d$ is the same as for the thinner layer, so you might expect everything to scale in the same way as d is changed. Thus t_s is not the actual time, but it is a useful *timescale* for this problem.

This expectation is borne out by a rigorous mathematical analysis, which yields an actual cooling time of $d^2/4\kappa$ [1]. As expected, the actual time is less than our upper bound, but it scales in the same way, so the actual time is always half of our upper bound. The temperature profile that results from a rigorous analysis is sketched in Figure 5.5. Mathematically, the curve is an *error function*, which I will spare you the details of. However, you will notice that the temperature approaches the initial temperature asymptotically with depth. This makes it harder to measure the thickness of the cooled layer. So as to be specific, the depth d is defined as the depth at which the temperature reaches 84% of the asymptotic temperature. (The 84% makes simple sense in the context of the error function.)

The result we have obtained is quite powerful. Let us write it in two different forms:

$$t \sim d^2/\kappa, \tag{5.15a}$$
$$d \sim (\kappa t)^{1/2}. \tag{5.15b}$$

The symbol '$\sim$' means 'is of the order of', or 'is roughly equal to'. Thus if we know a length scale in a thermal diffusion problem, we can simply estimate a corresponding timescale, as we have done above. You can as easily apply Eq. (5.15a) to a cooling magmatic sill or dyke, to a cooling pluton, or to the cooling crust after a thermo-tectonic event, for example.

Conversely, if we know a timescale, we can estimate a length scale. For example, old sea floor is about 100 Myr old. You can use Eq. (5.15b) to deduce that the oceanic lithosphere will have cooled to a depth of 57 km. The rigorous solution yields twice that, 114 km, close to the observed thickness of older oceanic lithosphere.

In this section we have covered heat conduction, and then considered how a temperature profile will change with time. This introduced the *thermal diffusivity*, and led to the simple relationship given in two forms in Eq. (5.15). We can use this directly in our extended theory of convection in the next section.

Diffusion occurs in other contexts, notably chemical diffusion, in which chemical species move along concentration gradients. Similar relationships apply, involving a chemical diffusivity, which is usually much smaller than thermal diffusivity. Thus the concepts and relationships developed here have even wider applicability.

5.5 Plate velocity – thermo-viscous version

We can now return to the calculation of plate velocity. In Section 5.3 we got as far as the following formula (Eq. (5.9)):

$$v = g\rho_0 \alpha T D d / 12\mu.$$

The depth d is the thickness of the plate at the surface, which we are identifying as the top thermal boundary layer, and which forms as follows (refer to Figure 5.2(a)). Hot mantle rises to the surface at a mid-ocean ridge spreading centre. It then turns and moves horizontally away from the spreading centre. As it drifts away it cools from the top by conduction, because it is in contact with cold sea water. The cooled layer thickens as time passes, until it reaches a subduction zone and is returned to the mantle. The thickness of the cooled layer, which we can recognise as a thermal boundary layer, depends on the time, t_p, for which the plate is at the Earth's surface. This time depends on how fast the plate goes and how far it has to go. In the simple box model of Figure 5.2(b) the distance travelled is D. If the plate velocity is v, then

$$t_p = D/v. \tag{5.16}$$

This time can now be used in (5.15b) to calculate the thickness d. I will use the result from the rigorous analysis, so

$$d = (4\kappa t_p)^{1/2} = (4\kappa D/v)^{1/2}. \tag{5.17}$$

This expression can now be substituted into the above formula for v. The only catch is that v now occurs on the right-hand side as well, and v is the unknown we are

5.5 Plate velocity – thermo-viscous version

trying to calculate. The situation can be retrieved with a bit of algebra. Squaring both sides yields

$$v^2 = (g\rho_0 \alpha T D/12\mu)^2 (4\kappa D/v),$$

so

$$v^3 = (g\rho_0 \alpha T D/12\mu)^2 (4\kappa D) = D^3 (g\rho_0 \alpha T 2\kappa^{1/2}/12\mu)^2$$

and finally

$$v = D \left(\frac{g\rho_0 \alpha T \sqrt{\kappa}}{6\mu} \right)^{2/3}. \tag{5.18}$$

This expression involves only g, the temperature and some material properties. Let's try evaluating it with $T = 1300\,°C$ and other quantities as used before. This gives $v = 3.6 \times 10^{-9}$ m/s, which is 0.115 m/yr or 11.5 cm/yr. Thus our extended theory yields a result within a factor of 2 of observed plate velocities, which are around 5 cm/yr.

It is time to review this chapter. We have reproduced the observed velocities of tectonic plates using a theory that assumes plate motions are governed by a balance between the (thermal, negative) buoyancy of a sinking plate and the viscous resistance of the fluid mantle. The theory, in its extended form, also assumes that plates are a cool thermal boundary layer formed by conductive cooling to the Earth's surface. The only inputs have been the inferred temperature and viscosity of the mantle, and some other material properties established by experiments. Thus we have used a simple theory of convection to predict the velocities of tectonic plates, and have achieved an agreement that is within the expected uncertainties of the approximate theory and the uncertainties of the inputs. We now have a possible answer to the question: 'What drives the plates?' The answer is that the plates drive themselves, because they are the active part of a form of mantle convection.

Does this 'prove' that this is why plates move? No, there might be a theory involving other forces that can also match the observations. But it does show that our simple theory *can* explain the observations, so it is a candidate for a good theory of why plates move. Finally, I hope there is no mystery in how we got to this point. We examined viscous fluid flow and we examined thermal expansion and thermal buoyancy. Then we put those things together to deduce plate velocities, using observational or experimental inputs to evaluate the required terms. Convection can be understood in fairly simple physical terms. It is not just something mysterious that happens 'down there'.

In the interpretation made here, tectonic plates do not ride passively on something mysteriously overturning down there. Tectonic plates are the active components in a form of mantle convection – they are the cool thermal boundary layer of the mantle convection system. Thus the relationship between plates and mantle convection is revealed, and it is a straightforward relationship, once we have a clear understanding of what convection is.

We can say this in another way. The cycle, in which mantle material rises at a mid-ocean ridge, cools at the surface to form a plate, moves horizontally and subducts, and finally is warmed and reabsorbed into the mantle, is a form of convection. The motion is driven by thermal buoyancy within the system, and it transports heat, as we will see explicitly a little further on.

We have established the central result of this book. The rest is tidying up and looking at some implications. Convection has been, to many people, a rather mysterious process. I hope this treatment has made convection a lot less mysterious. It is also widely regarded as complicated, requiring supercomputers to model. The simple theory here shows that, when approached with clear concepts and judicious approximations, convection is not so complicated, and in fact good quantitative agreement with observations can be obtained relatively easily.

5.6 The Rayleigh number and other fluid-dynamical beasts

Fluid dynamicists love to find dimensionless numbers that characterise fluid flow. They then name them after each other. For example, you can't read anything about convection without encountering the *Rayleigh number*. There will usually be a complicated-looking expression for it and, possibly but not necessarily, a terse explanation that you may or may not understand. Well, I'm being a little facetious and a little unkind, though the explanations usually are terse, or absent.

We have been through a very nice theory of convection without needing to mention the Rayleigh number, because I was more concerned with making the physics clear. However, to give fluid dynamicists their due credit, the Rayleigh number is extremely useful, because it gives a very concise summary of the convective system that allows you to know whether the convection is languid or very vigorous, or whether the convection in a laboratory tank is in any way similar to convection in the ocean or in the mantle.

Here is one way to get at the Rayleigh number. Use Eq. (5.18) in (5.17) to get an expression for d that doesn't involve v. Equation (5.17) is

$$d = (4\kappa D/v)^{1/2}.$$

Square both sides and substitute for v from (5.18):

$$d^2 = \frac{4\kappa D}{D} \left(\frac{6\mu}{g\rho_0 \alpha T \sqrt{\kappa}} \right)^{2/3}.$$

5.6 Rayleigh number and other fluid-dynamical beasts

Now take the power 3/2 on both sides, and rearrange a little:

$$d^3 = 8\kappa \left(\frac{6\mu}{g\rho_0 aT} \right).$$

Now divide by D^3 and tidy up:

$$\left(\frac{d}{D} \right)^3 = \frac{48\kappa\mu}{g\rho_0 \alpha T D^3}.$$

The left-hand side is a ratio of lengths, and therefore dimensionless. This means that the right-hand side must be dimensionless. The Rayleigh number is defined as the inverse of this collection of quantities, leaving out the numerical factor. In other words

$$\text{Ra} \equiv \frac{g\rho_0 aT D^3}{\kappa\mu}. \tag{5.19}$$

The Rayleigh number is generally symbolised as Ra, and the 'a' is not a subscript. We'll encounter other numbers with two-letter symbols shortly. The triple-bar equality means that this is a definition: the left-hand side equals the right-hand side by definition. The Rayleigh number is a dimensionless quantity, as already implied. Its full usefulness will become apparent after we define a couple of other numbers. First, however, note that the previous expression can now be written in the compact form

$$\frac{d}{D} = \left(\frac{\text{Ra}}{48} \right)^{-1/3}. \tag{5.20}$$

Now let's look at velocity again. The thermal diffusivity κ has dimensions of length^2/time. Therefore κ/D will have dimensions of velocity (length/time). We will use this as a velocity scale: $V = \kappa/D$. Now you can use Eq. (5.18) to show, with a little manipulation, that

$$\frac{v}{V} = \left(\frac{\text{Ra}}{6} \right)^{2/3}, \tag{5.21}$$

so we can also write the convective velocity in very compact form.

Finally let's look at heat flow. If the thermal boundary layer has a thickness of d, then, from Fourier's 'law' (Eq. (5.10) in Section 5.4), the heat flux through the boundary layer will be

$$q = KT/d.$$

This is also the total heat flux transported through the fluid, as it is the heat emerging at the surface. If the heat had to be conducted through the fluid layer, in other words if there were no convection, then the conducted heat flow, q_c, would be

$$q_c = KT/D.$$

The ratio of these quantities is just D/d, so from Eq. (5.20)

$$\frac{q}{q_c} = \left(\frac{\text{Ra}}{48}\right)^{1/3}. \tag{5.22}$$

Thus we have another very compact expression.

You can see that, apart from numerical factors, the Rayleigh number contains within it information about the boundary layer thickness, the velocity of convection and the amount of heat transported by convection. For example, the larger is Ra, the faster the convection will flow and the more heat it will transport. Perhaps less obviously, Eq. (5.20) shows that the more vigorous convection will have a thinner thermal boundary layer. That's because the faster fluid spends less time at the surface and does not cool to as great a depth.

Two of the ratios we have just defined also have names. The ratio of velocities in Eq. (5.21) is called the Peclet number, written Pe. The ratio of heat fluxes in Eq. (5.22) is called the Nusselt number, written Nu. The ratio of thicknesses in Eq. (5.20) has not been given a name. Thus we can summarise the above results as

$$d/D \sim \text{Ra}^{-1/3},$$
$$\text{Pe} = v/V \sim \text{Ra}^{2/3}, \tag{5.23}$$
$$\text{Nu} = q/q_c \sim \text{Ra}^{1/3}.$$

With the values we have been using, the mantle value of Ra is 3.5×10^6. This is regarded as a moderately high value, indicating reasonably vigorous convection. This qualitative statement is based on the fact that there will not be any convection unless Ra exceeds about 1000. We will look at this some more in Chapter 7, in the context of mantle plumes. Suffice to say for now that at low Ra any upwelling that begins to form, like that in Figure 5.1, is smoothed out by thermal diffusion before it can become large enough to rise significantly. The disturbance in the fluid dies away and the heat is transported by conduction through static fluid. So the mantle value of Ra is well above the level at which convection can begin.

Suppose you want to set up a laboratory experiment to model convection in the mantle. How do you know what tank size and fluid properties to use? Well, if the tank experiment has a similar Rayleigh number to the mantle, then the flows should have a similar vigour. Suppose your laboratory fluid has a density similar to water, 1000 kg/m^3, thermal diffusivity $10^{-7} \text{ m}^2/\text{s}$, thermal expansion coefficient $5 \times 10^{-4}\,°\text{C}^{-1}$ and viscosity 5 Pa s. If you can impose a temperature difference of

50 °C, how deep does your tank have to be to have the same Ra as the mantle? The answer is 19 cm. This is a feasible scale for a laboratory experiment, and the vigour of your experiment will be analogous to the vigour of the mantle.

However, having the same value of Ra doesn't ensure that everything about the experiment will be like mantle convection. For example, it is difficult to get a laboratory material that will give a good simulation of tectonic plates. The presence of plates as part of mantle convection has a big effect on the form of the convective flow, and on some of the main things we can observe about mantle convection, as we will see in the next chapter.

6
The plate mode of convection

> Mechanical properties change with temperature, from brittle plate to yielding mantle and back. This strongly affects their dynamical behaviour and their influence on convection. Plates organise the flow. Internal heating versus bottom heating also affects the form of convection. The plate cycle (formation, cooling, subduction, reabsorption) is convection. The plate mode of mantle convection transports a large fraction of Earth's heat budget. Seafloor topography and heat flow can be quantitatively explained with remarkable success.

The convection theory developed in the previous chapter applies to many forms of convection, and it seems to apply reasonably well to mantle convection, but with some important qualifications. Mantle convection takes distinctive forms that in some ways are quite unlike familiar examples of convection such as occur in familiar kinds of fluid. The main reason for the differences is that the mechanical behaviour of mantle rocks changes quite dramatically between the temperature at the Earth's surface and the temperature within the mantle.

6.1 The strong lithosphere

The temperature dependence of viscosity shown in Figure 4.4 tells us that reducing the temperature from $1300\,°C$ to $1000\,°C$ will increase the viscosity of mantle rocks by as much as three orders of magnitude – a factor of 1000. However, if the mantle rocks are much cooler than that, they cease to deform like a viscous fluid. Through an intermediate range of temperature they develop ductile shear zones, so that the deformation is concentrated in relatively narrow zones instead of occurring uniformly through the fluid. At temperatures lower than a few hundred degrees,

6.1 The strong lithosphere

the shear zones become faults, and the material is better described as a brittle solid rather than as a fluid.

The general term for the mechanical response of materials to forces and stresses is *rheology*, and the rheology of rocks is quite complicated, depending not only on temperature but also on such things as the confining pressure, the shear stress causing deformation, the particular minerals present, their grain size and some details of their chemical composition, such as a few hundred parts per million of dissolved water. It is thus hard to characterise the rheology of mantle rocks, and they are not all that well known in the intermediate ranges of temperature and stress that apply in the Earth's lithosphere. Nevertheless a couple of general points can be made fairly confidently.

At the Earth's surface, rocks are brittle. They will break along fractures, and sliding can occur along fractures, which are then known as faults. At increasing depths, the stress required to cause sliding increases because friction increases with increasing confining pressure. On the other hand, at a sufficiently high temperature the rock will deform like a fluid. If the fluid-like deformation is fast enough, it may relieve the stress before any more fracturing or sliding occurs, and fluid deformation will then be the dominant mode of deformation. At greater depths and temperatures, the rocks will deform more easily, so their resistance to deformation will decrease again.

This general pattern is illustrated in Figure 6.1 for oceanic lithosphere and for continental lithosphere. It shows a plot of 'strength', loosely defined, versus depth. For oceanic lithosphere the strength increases with depth, starting from the surface. Temperature also increases with depth through oceanic lithosphere, as depicted schematically in Figure 5.5. The increasing temperature promotes fluid-like deformation, so that at depths greater than about 40 km the strength decreases with increasing depth as the fluid deformation becomes dominant. In this example there is an intermediate regime called semi-brittle deformation that limits the strength between about 10 km and 40 km depth. The three deformation regimes (brittle, semi-brittle and fluid/ductile) define a *strength envelope*, which indicates roughly what level of shear stress the rocks can resist at each depth.

The strength envelope for continents is rather different, because crustal rocks are generally not as strong as mantle rocks. Thus the fluid-like deformation becomes dominant at about 15 km depth in this example, and the lower crust is relatively deformable. In this example the crust is taken to be 35 km thick, and below this depth the strength jumps up again because of the greater strength of mantle rocks. Thus the continental lithosphere may be stronger near the top and bottom than in the middle, within the lower crust.

Because its middle part is relatively weak, the continental lithosphere is also weaker overall than oceanic lithosphere. This inference, based on laboratory

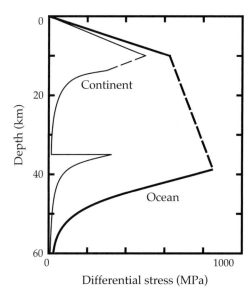

Figure 6.1. Strength envelopes estimated for representative oceanic and continental geotherms. Such estimates depend greatly on details assumed (see text). Each case comprises three regimes: brittle (straight lines), semi-brittle (dashed lines) and fluid/ductile (curves). In the continental case, the crustal ductile response changes to the mantle ductile response at 35 km depth. In these examples the deformation rate is assumed to be 10^{-15} s^{-1}. After Kohlstedt *et al.* [39]. Copyright by the American Geophysical Union.

measurements of rock deformation, is borne out by the fact that the continental parts of tectonic plates have more internal deformation, as exemplified by the broad deformation zones through China and the western USA. In contrast, oceanic lithosphere undergoes little deformation away from plate boundaries.

As we saw in Chapter 3, Tuzo Wilson conceived of plate tectonics by looking at the surface tectonic features of the Earth. What he saw was a network of mobile belts dividing the Earth's surface into large pieces, like a jigsaw puzzle. He named the pieces *plates*, and described them as essentially rigid, though moving. The fact that the Earth's surface is not deforming in most places, and that deformation is usually confined to narrow zones, implies that the lithosphere behaves like a brittle solid, to a good approximation. This is why Wilson was able to take the notion of a mobile belt to its geometric limit and treat it as a narrow fault. He also took the behaviour of the intervening plates to its limit, and described them as rigid, even though there is some relatively small deformation in some parts of some plates.

The implication is that the behaviour of the lithosphere is dominated by its upper, brittle part. Its lower parts evidently adjust to the pattern of deformation imposed by the upper brittle part, even though they may be semi-brittle or 'ductile', in the sense of forming ductile shear zones. (Strictly speaking, ductile refers to the ability of a material to be stretched out. Rocks have little ductility in this sense, because they have very low strength in extension. It is better to describe them as malleable or, as I have been doing, as undergoing fluid-like deformation.) Thus the major fault boundaries separating the plates may be literally faults at the surface, but grading into ductile shear zones and broader deformation zones at increasing depth. Away from plate boundaries, the strength of the upper brittle part evidently shields the lower, more deformable parts from undergoing much deformation.

6.2 The role of the lithosphere in mantle convection

The lithosphere is strong because it is cold. Low temperatures (and relatively low pressures) are required for brittle behaviour to occur. At the temperature of the interior of the mantle (1300 °C or more), the mantle rocks can flow fast enough to relieve the stresses driving convection, and they are therefore relatively weak.

The lithosphere is cold because heat is conducting to the Earth's surface, which is cold. In the previous chapter we saw that within 100 Myr, a typical age of old lithosphere, the mantle cools to a depth of about 100 km. We also saw that the depth to which the mantle has cooled is proportional to the square root of the time it has been cooling.

Because the lithosphere is cold, it is negatively buoyant, meaning it is denser and heavier than the underlying hot mantle. It is the negative buoyancy of the lithosphere that causes it to sink into the mantle, pulling the surface lithosphere along behind it and pushing the mantle around as it sinks. In other words, it is the negative buoyancy of the lithosphere that drives mantle convection.

Putting these things together, we can recognise the lithosphere as the cool thermal boundary layer at the top of the mantle fluid that drives convection in the mantle. We saw in the previous chapter that convection is driven by the buoyancy of thermal boundary layers. In this view, then, the cool thermal boundary layer forms as hot mantle rises to the surface (refer to Figure 5.2), where it loses heat by conduction to the cold surface. As the material moves horizontally along at the surface, the thermal boundary layer thickens, until at some point it sinks back into the mantle under the action of its own excess weight, or negative buoyancy.

At what point will the lithosphere sink back into the mantle? This question leads us to a key distinction between mantle convection and more familiar kinds of convection. In 'normal' convection, the thermal boundary layer becomes unstable and 'drips' away into the interior of the fluid, as illustrated in Figure 6.2. This behaviour depends on the fluid in the thermal boundary layer being able to deform, so it can form a drip. For most familiar fluids the viscosity does not change with temperature, so the fluid in the cool thermal boundary layer is just as mobile as the fluid in the interior.

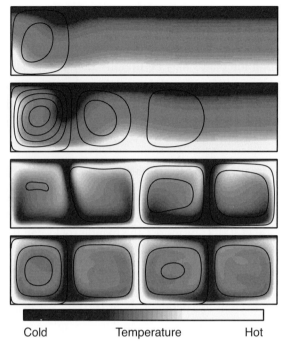

Figure 6.2. Convection beginning in a 'normal' fluid, i.e. one whose viscosity is constant. In this example there are two thermal boundary layers, a cool one at the top and a hot one at the bottom (where heat is conducting into the fluid). As the cool thermal boundary layer thickens, it becomes unstable and forms a 'drip' that sinks into the interior of the fluid. The lower thermal boundary layer also becomes unstable and forms a rising blob at the left. The initial instabilities then trigger a chain reaction in which each thermal boundary layer forms a succession of drips that sink or rise through the fluid, until a set of convection cells is formed.

In the mantle, on the other hand, the lithosphere is so strong that it can prevent the formation of a drip. It is possible this could inhibit convection, or at least prevent the lithospere from deforming at all. This is the situation with Mars and the Moon, whose lithospheres have not deformed since early in their history. Their lithospheres are undeformed and unbroken by major faults, and there is no plate tectonics operating on either body. They are sometimes also called 'one-plate' planets – you can think of the unbroken lithosphere as being a single plate that embraces the entire planet.

Evidently on Earth there are stresses large enough to overcome the strength of the lithosphere and to break it into pieces. It is evidently only at plate margins that the lithosphere is weak enough to allow parts of it to sink. The result is what we call a subduction zone (sketched in Figure 5.2). It is also only at plate margins that upwelling mantle reaches the surface, at a mid-ocean ridge or 'spreading centre'. Occasionally a plate may break and form new plate margins, either subducting or

6.2 The role of the lithosphere in mantle convection

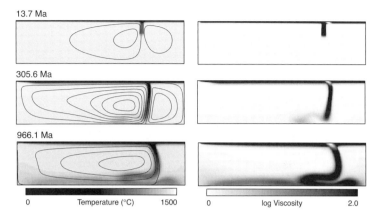

Figure 6.3. Convection beginning in a fluid with a stiff upper thermal boundary layer. The left panels show temperature, in grey scale, and streamlines. The right panels show the viscosity, relative to the least viscous fluid, in a logarithmic grey scale. The viscosity depends on temperature, though it is capped at a maximum of 100. It is also reduced, artificially, at either end of the left-hand plate.

spreading, but while the plate is intact there is no upwelling or downwelling that involves the surface.

This inhibition of deformation within the lithosphere has a major effect on the spatial *pattern* of convection in the mantle. It means that the plates determine where the upwellings and downwellings of convection will be. This effect is illustrated in Figure 6.3, which shows convection beginning in a fluid whose viscosity depends on temperature, so that the viscosity in the thermal boundary layer is higher than in the interior of the fluid. However, the viscosity is prescribed to be lower at two places along the top boundary. This has the effect of separating the stiff thermal boundary layer into two pieces, simulating plates. The left-hand plate is started with a cold segment hanging down, so as to initiate flow. Subsequently the left-hand plate continues to subduct at this location, and new fluid wells up at the left side of the box and cools as it moves to the right, thus adding to and maintaining the surface plate. The subducted portion of the plate sinks through the fluid and folds onto the bottom of the box. Its viscosity stays relatively high for quite a long time in this numerical model. (The convection in this model is less vigorous than actual mantle convection, which is why the times shown on the boxes are quite long.)

The important point illustrated by Figure 6.3 is that the locations of the upwelling and downwelling remain confined to the edges of the plate. No new drips form in the middle of the left-hand plate, because the fluid within the plate is too stiff. This contrasts with the behaviour shown in Figure 6.2, in which the distance between upwellings and downwellings is similar to the depth of the box. If the viscosity of the thermal boundary layer was the same as that in the interior in Figure 6.3,

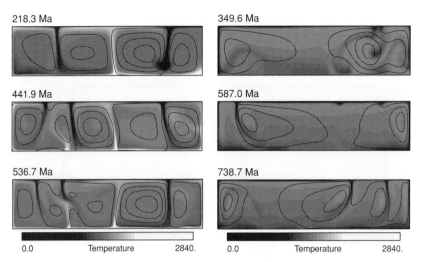

Figure 6.4. Convection in a constant-viscosity fluid. In the left-hand panels the fluid is heated through the base and cooled through the top, so there is a hot thermal boundary layer at the bottom and a cool thermal boundary layer at the top. In the right-hand panels the fluid is heated internally and cooled through the top, so there is only a cool thermal boundary layer at the top, from which drips fall erratically. (Internal heating simulates the effect of radioactive heating of the mantle.)

drips would form well before the plate material reached the subduction zone. In Figure 6.3, the convection 'cell' under the left-hand plate is about three times wider than the depth of the box. This corresponds roughly with the situation in the mantle. The mantle is about 3000 km deep, yet the larger plates are 7000 to 14 000 km across, two to four times the depth of the mantle. The ratio of cell width to depth is often called the *aspect ratio*, and for the larger plates the aspect ratio is 2–4.

Actually the examples in Figures 6.2 and 6.3 are not strictly comparable because in Figure 6.2 the vigour of convection is lower and it has a hot thermal boundary layer at the bottom. However, the distinctly different behaviour in Figure 6.3 is not due to either of these factors. This can be seen from Figure 6.4, which shows convection in a fluid of constant viscosity at a vigour similar to that of Figure 6.3. The left-hand panels have both thermal boundary layers (in other words, the fluid is heated through the base as well as cooled through the top), whereas the right-hand panels have only the top thermal boundary layer, as in Figure 6.3. As you can see, the pattern of convection in Figure 6.4, right-hand panels, is quite different from that in Figure 6.3. In Figure 6.4, drips form erratically at the top surface, and they actually migrate laterally with time, which is why they are not vertical columns.

It is worth mentioning that two other factors may contribute to the large aspect ratio of mantle convection [1]. One is that the lower mantle is inferred to have a viscosity roughly 10 to 100 times that of the upper mantle [41]. This tends to favour large-scale flow because then the rates of deformation involved are smaller for a given velocity, and so the viscous resistance is reduced. The other factor is the spherical geometry of the mantle. The mantle extends about half-way to the centre of the Earth, which means that the deep return flow has to travel only about half as far as a surface plate. There will therefore be less time for an instability to develop and rise under the middle of a plate.

There is another significant distinction between convection with plates and 'normal' convection. It is that only one plate subducts, so the downwelling is completely asymmetric (Figure 6.3, bottom left panel). In constant-viscosity convection, on the other hand, the downwelling is more symmetric, drawing fluid from both sides of the downwelling.

To summarise this section, the presence of plates has a large effect on the spatial pattern of convection. Upwellings and downwellings occur only at plate margins, and not, so far as we can see, in between. In this sense the plates *organise* the pattern of the flow. As a result, plates can be much wider than the depth of the mantle. Also as a result, subduction is asymmetric, with material from only one side descending into the mantle.

The distinctive form of mantle convection occurs because the mantle material changes from acting like a viscous fluid to acting like a brittle solid, and back, as it rises to the surface, cools and then sinks back into the mantle. It is the brittleness of the material at the surface that gives rise to plate tectonics. This is also the reason it was so hard for people to recognise the plates as part of a convecting system. The plates have a range of sizes, strange shapes and in many places their boundaries have sharp corners. None of the latter features look like the top of a 'normal' convecting fluid.

6.3 Heat transport – the plates are mantle convection

The further implication of this picture is that the plates are *part of* the convecting system. They comprise one of the driving thermal boundary layers of mantle convection. They do not ride passively on a convection system that occurs mysteriously under the plates for unspecified reasons. Nor is the cycle of upwelling under a spreading centre, cooling at the surface and sinking back into the mantle to be reheated something that is independent of a convecting system. This cycle *is* the convecting system in action. It is the process by which mantle convection removes heat from the deep interior of the mantle.

I have called the convection involving plates the *plate mode* of mantle convection [1]. It deserves its own name because there is another mode of mantle convection with a rather different form, the plume mode, that is driven by the hot thermal boundary layer at the bottom of the mantle, as will be discussed in the next chapter.

We can quantify the rate at which heat is removed from the mantle by the formation and subduction of plates. Referring to Figure 5.4(b) showing schematic profiles of temperature through a thermal boundary layer, mantle that rises under a mid-ocean ridge has the mantle temperature, $T_m = 1300\,°C$, essentially right to the surface. When this layer of mantle reaches a subduction zone about 100 Myr later, it has cooled to a depth of about 100 km. Its temperature then varies from $T_0 = 0\,°C$ at the surface to $1300\,°C$ at 100 km depth. Its *average* temperature is then about $650\,°C$. When it started at the ridge, its average temperature was $1300\,°C$. Therefore its average temperature has dropped from $1300\,°C$ to $650\,°C$, a *decrease* of $650\,°C$. Writing this out, the change in the average temperature, ΔT, is

$$\begin{aligned}\Delta T &= T_{av} - T_{av}^i \\ &= (T_0 + T_m)/2 - T_m \\ &= (T_0 - T_m)/2 \\ \Delta T &= -T_m/2,\end{aligned} \quad (6.1)$$

where T_{av}^i is the initial average temperature at the ridge.

Referring back to Eq. (5.12) and the paragraph preceding it, we can write an expression for the resulting change in the heat content of the lithosphere. A vertical column of rock with a cross-sectional area of $1\,m^2$ and extending to a depth d has a volume of $d \times 1 \times 1$ and a mass of ρd, where ρ is the density of the rock. Therefore the change in heat content, ΔH, of this rock column is

$$\Delta H = \rho d C_P \Delta T, \quad (6.2)$$

where C_P is the specific heat of the rock.

Seafloor spreading creates an area of about $3\,km^2$ of new sea floor every year [42]. Subduction removes a similar amount. Remembering that there are about 3×10^7 seconds in a year, and converting the area to square metres, the rate of removal of sea floor by subduction is $S = 3 \times 10^6 / 3 \times 10^7\,m^2/s = 0.1\,m^2/s$.

Every square metre of sea floor that is removed by subduction has lost an amount of heat ΔH given by Eq. (6.2). Therefore the total rate of heat loss by cooling of the oceanic lithosphere is

$$\begin{aligned}Q &= S\Delta H \\ &= S\rho d C_P T_m / 2.\end{aligned} \quad (6.3)$$

Using $\rho = 3300\,kg/m^3$ and $C_P = 1000\,J/kg\,°C$, this yields $Q = 2.1 \times 10^{13}\,W = 21\,TW$.

Table 6.1. *Contributions to global heat flow.*

	Area (10^8 km^2)	Mean heat flux mW/m^2	Total heat flow (TW)	Percentage of global
1. Sea floor	3.1	100	31	76
2. Continental crust	2.0	50	10	24
a. Crustal radiogenic	–	25	5	12
b. Mantle	–	25	5	12
3. Total mantle (1 + 2b)	5.1	70	36	88
4. Total global (1 + 2a + 2b)	5.1	80	41	100

Another way to get to this result is to calculate the average heat flux, q, required to release the heat ΔH over the time, $t_s = 100$ Myr, that the plate material takes to reach a subduction zone:

$$q = \Delta H / t_s. \tag{6.4}$$

Using the above values, $\Delta H = 2.1 \times 10^{14}$ J/m^2 and $t_s = 3 \times 10^{15}$ s, so $q = 70$ mW/m^2. The surface area of the sea floor is 3.1×10^8 km^2, so the total heat loss through the sea floor is $Q = 22$ TW. This is essentially the same answer as obtained from Eq. (6.3), apart from rounding errors, as it should be because the two approaches are equivalent.

This rough estimate is a little lower than what is observed. The average heat flux through the sea floor is measured to be about 100 mW/m^2, which gives a total heat flow of 31 TW [11]. A more sophisticated mathematical analysis will get closer to the observed value. The point of our rough estimate is, as usual, to keep the physics clearly in view. Thus we can conclude that the cycle of plate formation and subduction removes around 30 TW of heat from the mantle.

We can compare this estimate with the total heat loss of the Earth, and the total heat loss of the mantle. The total heat flow out of the Earth is 41 TW [11]. Some of this heat is generated in the upper continental crust, and so is not part of the mantle heat budget. The continents have an average heat flux of about 50 mW/m^2 and an area of 2×10^8 km^2, so the total heat flow out of continents is about 10 TW. About half of this is generated by the radioactivity concentrated in the upper 10 km or so of the continental crust [1], and this heat conducts directly to the surface and so plays no part in driving mantle convection. The other half of the continental heat flow must then come from the mantle, conducting into the base of the continental lithosphere. Thus about half of the total continental heat flux, 25 mW/m^2, is emerging from the mantle through the continents, or about 5 TW in total.

These relationships are summarised in Table 6.1. About 75% of the Earth's heat loss occurs through the sea floor as a result of plate-scale convection. This is nearly

90% of the total heat loss from the mantle, the balance being lost by conduction into and through the continents.

The plate-scale flow is thus the dominant means by which heat is lost from the mantle. The balance of mantle heat loss can be accounted for by conduction through continental lithosphere. Thus we can conclude that, by the fundamental criterion of the amount of heat removed from the mantle, the plate-scale flow is the dominant mode of convection driven by the top thermal boundary layer of the mantle.

In other words it is not necessary to invoke any other mode of convection in order to explain the heat flowing out of the mantle. So-called small-scale convection used to be frequently discussed, and one form of this would, for example, be small-scale drips falling from the lowest, softest part of the lithosphere. We will look at such alternatives in a later chapter, but already we can conclude that any such additional mode of convection must not be transporting much heat, because we have explained the observed heat flow by invoking just the plate mode of convection. Plumes have also sometimes been invoked as an alternative or additional mode, but plumes are not driven by the top thermal boundary layer, and their role is not to remove heat from the mantle but to bring heat into the mantle, as we will see in the next chapter. Thus plumes may be an additional mode of mantle convection, but they are not an alternative to the plate mode.

The argument that most of the Earth's heat loss can be accounted for by the plate mode of convection is strengthened if we look at the geographic distribution of heat flow, and its close associate, topography. We will now look at these in reverse order.

6.4 The geography of topography

Chapter 2 summarised the observations that the depth of the sea floor increases in proportion to the square root of its age (Figure 2.5) and the heat flux through the sea floor varies inversely with the square root of its age (Figure 2.6). The description we have developed of the oceanic lithosphere as a thermal boundary layer can be used to derive the variation of these quantities that is implied by our theory. Both can be obtained from Eq. (5.13b), which relates the thickness, d, of the thermal boundary layer to the time, t, for which it has been cooling. It can be rewritten here as $d = \sqrt{2\kappa t}$. That version was based on a simple approximation, and it is useful here to use a result from a more rigorous calculation [1]:

$$d = 4\sqrt{\frac{\kappa t}{\pi}}. \tag{6.5}$$

As the oceanic lithosphere cools and thickens, it will undergo thermal contraction. A column of mantle with unit area in cross-section and extending to depth

6.4 The geography of topography

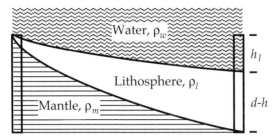

Figure 6.5. Subsidence of the sea floor due to thermal contraction.

d below a mid-ocean ridge will shrink as the lithosphere moves away, so that a similar column away from the ridge will occupy a height $d - h$, as illustrated in Figure 6.5. In other words the volume of a unit-area column changes from $d \times 1 \times 1\,\text{m}^2$ to $(d - h) \times 1 \times 1\,\text{m}^2$. The fractional change in volume is then h/d, and according to Eq. (5.2b) in our earlier discussion of thermal expansion

$$h/d = \alpha \Delta T \tag{6.6a}$$

or

$$h = \alpha \Delta T d, \tag{6.6b}$$

where α is the volume coefficient of thermal expansion and ΔT is the average change in temperature in the column. Equation (6.5) implies that h is proportional to the square root of the age of the sea floor. For an age of 100 Ma, using $\alpha = 3 \times 10^{-5}\,°\text{C}^{-1}$, $\Delta T = 650\,°\text{C}$, and $\kappa = 10^{-6}\,\text{m}^2/\text{s}$, the lithosphere thickness given by Eq. (6.5) is then 127 km, and the thermal contraction is $h = 2.5$ km. However, there is another consideration.

Equation (6.6b) would give the subsidence of the surface of the lithosphere if there was no ocean, but the presence of the water causes an isostatic adjustment: the weight of the water weighs the surface down further. Isostatic balance requires that there is no horizontal pressure gradient under the lithosphere, otherwise the material under the lithosphere would flow horizontally and that would raise one side and lower the other. We can take isostatic balance into account: the ideas are not hard to understand, though the book-keeping is a little messy.

To simplify our expressions, let's ignore the weight of the water above the level of the mid-ocean ridge crest, because that is the same for both columns and does not change the balance. We must also allow that the actual subsidence after isostatic balance has been established, h_I, is different from the subsidence just due to thermal contraction, h. Then the height of the right-hand column of rock plus water is $(d - h + h_I)$ and the pressure under it is

$$P_r = g[\rho_l(d - h) + \rho_w h_I], \tag{6.7}$$

where g is the acceleration due to gravity, ρ_l is the density of the lithosphere and ρ_w is the density of water. We must consider a left-hand column that extends to the same depth, so the pressure under the left-hand column is

$$P_l = g\rho_m(d - h + h_\mathrm{I}), \tag{6.8}$$

where ρ_m is the density of the mantle. If we equate these two pressures and rearrange we get

$$h_\mathrm{I}(\rho_m - \rho_w) = (\rho_l - \rho_m)(d - h). \tag{6.9}$$

This expression makes sense if you notice that the left-hand side is (minus) the mass deficit in the upper part of the right-hand column, relative to the left-hand column, and the right-hand side is the mass excess in the lower part, and these must sum to zero.

We have one more step to complete. The lithosphere is denser than the underlying mantle because of thermal contraction. Therefore we can use Eq. 5.3 to relate them:

$$\frac{\rho_l - \rho_m}{\rho_m} = \alpha \Delta T = \frac{h}{d}.$$

Substituting in Eq. (6.9) gives

$$h_\mathrm{I}(\rho_m - \rho_w) = h\rho_m(1 - h/d)$$

and because h/d is much less than 1 it can be neglected, so

$$h_\mathrm{I}(\rho_m - \rho_w) = h\rho_m$$

and we get our expression for the subsidence with isostatic balance:

$$h_\mathrm{I} = h\left(\frac{\rho_m}{\rho_m - \rho_w}\right). \tag{6.10}$$

Finally, to get the subsidence as a function of the age of the sea floor, we can substitute for h from Eq. (6.6b) and for d from Eq. (6.5), so we end up with

$$h_\mathrm{I} = \left(\frac{\rho_m}{\rho_m - \rho_w}\right)\alpha\Delta T 4\sqrt{\frac{\kappa t}{\pi}}. \tag{6.11}$$

Again, the subsidence is proportional to the square root of the age of the sea floor, but we have taken the trouble to account for isostasy, which changes the answer by the factor $\rho_m/(\rho_m - \rho_w)$. Using $\rho_m = 3300$ kg/m^3 and $\rho_w = 1000$ kg/m^3, this factor is 1.43. The isostatic subsidence is then 3.5 km. This is quite a good approximation to the observed subsidence of 100 Ma old sea floor, which is a little more than 3 km (Figure 2.5).

Thus both the functional form of seafloor subsidence (proportional to the square root of age) and the amplitude are well accounted for by our theory. This theory is, basically, that the oceanic lithosphere is a thermal boundary layer formed by conductive cooling at the Earth's surface.

6.5 The geography of heat flow

The same theory allows us to calculate how the flux of heat reaching the sea floor by conduction from depth should vary with seafloor age. In Section 5.4 we used a simple approximation, $q = KT_m/d$ (Eq. 5.11), based on Figure 5.4. Here K is the thermal conductivity, and the formula is an expression of Fourier's law of conduction, which was introduced in Section 5.4. Substituting from Eq. (6.5) for d shows that this simple formula implies that q is inversely proportional to the square root of age:

$$q = \frac{KT_m}{4}\sqrt{\frac{\pi}{\kappa t}}.$$

However, again, here it is useful to use a more accurate formula, based on the rigorous error-function solution for the temperature profile through the oceanic lithosphere [1]. This result is

$$q = \frac{KT_m}{\sqrt{\pi \kappa t}}. \tag{6.12}$$

Using $K = 3$ W/m °C and other values as above, this gives a heat flow of 40 mW/m² at an age of 100 Ma. This is a good approximation to the observed magnitude of heat flow through old sea floor (Figure 2.6). Again, treating the oceanic lithosphere as a thermal boundary layer yields good agreement with both the functional form of the dependence of heat flow on age and its magnitude.

6.6 Numerical model of the plate mode

The results just calculated assume that the deeper mantle has no significant effect on the topography and heat flow, because we have only looked at the cooling thermal boundary layer at the surface, and have not considered the effect of the boundary layer subducting and flowing around the mantle. Figure 6.6 shows that a numerical convection model that includes the subducted lithosphere still yields similar topography. The topography has a slightly smaller amplitude in this model than the equivalent simple boundary layer theory, shown dashed, and the form deviates a little from the square root of age profile. Nevertheless, the numerical topography is similar within the scatter of the data in Figure 2.5.

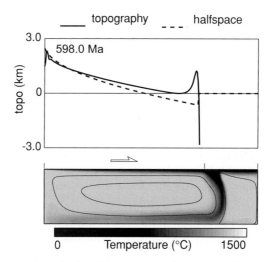

Figure 6.6. Topography calculated from a numerical convection model. The model is not accurate for subduction zones. The dashed 'half-space' line refers to the thermal boundary layer theory.

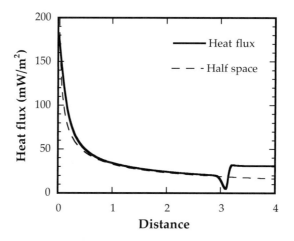

Figure 6.7. Surface heat flux calculated from the numerical convection model of Figure 6.6. The dashed 'half-space' line refers to the thermal boundary layer theory.

The surface heat flow from this model is shown in Figure 6.7. In this case the magnitude and general form are quite close to that of the thermal boundary layer theory. This is to be expected because the surface heat flow is not sensitive to the deep mantle, but only to the temperature variation within the thermal boundary layer.

6.7 Summary of the plate mode

In Chapter 5 we developed a simple theory of mantle convection and showed that it was capable of predicting the velocities of plates to quite good accuracy, given our initial ignorance and the simplicity of the ideas and approximations used. In this chapter we have developed this picture by taking account of the distinctive mechanical properties of the lithosphere, and deduced some important results.

Mantle material changes its mechanical properties quite radically as it rises from the deep, hot interior, where it behaves like a viscous fluid, and cools at the surface, becoming strong and effectively brittle. Mantle convection takes a distinctive form because of this change. The lithosphere is strong, but broken into pieces that we call plates. The plates control the spatial pattern of convection, because upwelling and downwelling occur only (or predominantly) at plate margins. The plates thus define what I have called the *plate mode* of mantle convection, with 'cells' that can be up to four times wider than the depth of the mantle. The appearance of this mode of convection at the Earth's surface is unusual because the convecting material is a brittle solid at the surface, and this results in plates having a range of sizes, odd shapes and in some places quite angular boundaries. This is part of the reason it was hard to recognise as convection, and why there has been a lot of confusion about the relationship between the plates and mantle convection.

In this picture, the plates are an integral part of mantle convection. In fact, they are the most active component, comprising the driving thermal boundary of the plate mode. The theory of thermal diffusion gives us a good account of the thickness of the oceanic plates. As a direct consequence, it also gives a good account of seafloor subsidence away from mid-ocean ridges, and of the variation of seafloor heat flux away from ridges. The plate mode accounts for most of the heat observed to be emerging from the Earth's interior.

This empirical success is obtained by considering only the top 100 km or so of the mantle, as we have done in the theory of the cooling, thickening plate. An important implication is that the underlying mantle must be relatively passive, otherwise it would disrupt the regular subsidence of the lithosphere. There are deviations from this regular behaviour, as we will see in Chapter 8, but they are secondary, and we can describe much of the important behaviour of the plate mode with this relatively simple theory.

There is more to remark about this empirical success. We noted in Chapter 2 that the mid-ocean ridge system is the second-largest topographic feature of the planet, after the continent–ocean dichotomy. A rather simple theory has successfully and quantitatively accounted for this topography, which we can recognise as having a dynamic origin – in other words, as resulting from the internal dynamics of the mantle.

A further important implication is that the plate mode must be the dominant mode of convection in the mantle, because any other mode would produce different patterns of topography and heat flow. The plate mode accounts for essentially all of the heat flow through the sea floor, so heat transport by any other mode must be secondary. We will see this explicitly in the next chapter, where we consider mantle plumes and their associated topography and heat flow. Plume-related topography is recognisable and distinctive, but clearly secondary compared with plate-related topography, implying that plume convection is secondary in the mantle, a subject that will be explored in the next chapter.

7

The plume mode of convection

> How were plumes conceived? Is there evidence for plumes at the Earth's surface? Hotspot tracks. Inevitable plumes? Hotspot swells, heat and mass transport by the plume mode. Degree of melting. Favoured forms of upwelling and the role of temperature dependence of viscosity. Heads and tails – flood basalts, connecting tracks. Melt production from plume heads, and relation to flood basalts. Irregularities, misfits, puzzles and possibilities – the simple thermal model doesn't explain everything. Thermochemical plumes may account for much of the irregularity.

Scattered across the sea floor are many submarine ridges, seamounts and plateaus (Figure 2.4). These are not obviously related to plate tectonics, as are the mid-ocean ridges and deep ocean trenches. Some of these edifices reach above sea level to form islands. In Chapter 3 I recounted how Tuzo Wilson built on the observations of Darwin and Dana, who had discerned an apparent age progression along island chains in the Pacific, from volcanically active islands through eroded islands to atolls marking where an island had been eroded to sea level. The classic example is the Hawaiian island chain, shown in Figure 7.1. As well as the islands, there is a long chain of seamounts extending to the northwest. There is also a broad swell in the sea floor around the Hawaiian chain. These observations give us important information about processes in the mantle.

Wilson noted that there are other such island chains around the world. Some are in the middle of plates, like Hawaii, whereas others are on or near mid-ocean ridges, like Iceland or Tristan da Cunha in the Atlantic. He proposed that the islands are generated volcanically from a source deep in the mantle, and that the age-progressive chain develops because the lithosphere moves over the source

74   *The plume mode of convection*

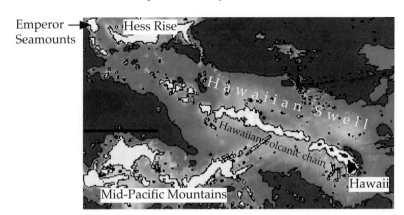

Figure 7.1. Topography of the sea floor near Hawaii, showing the Hawaiian seamount chain (white), the Hawaiian islands (black) and the Hawaiian swell. Contours are at depths of 3800 m and 5400 m.

[25]. Wilson called the mantle source a mantle 'hotspot', and he imagined it as a volume of hotter mantle located near the centre of a convection 'cell', which would account for it not moving with the lithosphere at the surface. Wilson's idea did not immediately catch on, perhaps because his idea of a hotspot in the middle of the mantle seemed *ad hoc*, and also because the age progression of the Hawaiian chain was only just being accurately established in 1963 [26, 27]. Nevertheless we can recognise Wilson's observations and ideas as a seminal contribution to understanding mantle convection.

In 1971 Jason Morgan proposed an alternative mechanism for the volcanism [43, 44]. Instead of a hot 'spot' somewhere in the middle of the mantle, Morgan proposed that a thin, hot column of mantle material rises from the core–mantle boundary. Such buoyant columns are known in other fluid-dynamical contexts (such as the smoke column rising from a candle), and he adopted the existing term *plume* for his proposed mantle columns. Morgan supposed that the heat carried by the plume comes from the core. This mechanism, though novel at the time, had the merit of being based on a plausible fluid-dynamical process and of having an energy source that could persist for the tens of millions of years required by some of the seamount chains. Thus, through Wilson's and Morgan's insights, the concept of a mantle plume was born.

The term 'hotspot' used by Wilson referred to a hot volume somewhere deep in the mantle, but its usage began to shift after Morgan's hypothesis. Wilson's version was quickly dropped, so it seems useful to use the term to refer to the volcanic centres at the Earth's surface. To avoid any confusion, one might speak of a *volcanic hotspot*. The chains of islands and seamounts extending away from active volcanic hotspots have become known as *hotspot tracks*.

Morgan proposed a second hypothesis, one that gained his ideas a lot of attention, though it is not required by the idea of a plume, and it has led to some confusion in later debates about plumes. Morgan proposed that volcanic hotspots do not move relative to each other. His idea was soon encapsulated in the term *fixed hotspots*. An important reason for this hypothesis gaining early attention was that it potentially provided a framework against which plate motions could be measured, the so-called *fixed hotspot reference frame*. Morgan had been one of the first to determine the relative motions of many of the plates [45], and he was actively interested in using the hotspots to refine plate motions, a project that was eagerly taken up by others as well. At the time it was still widely believed that the lower mantle had such a high viscosity that it would be essentially immobile, and this provided a plausible reason for hotspot 'fixity' (and Morgan actually argued that plumes represented a way in which the lower mantle could convect in spite of its stiffness). However, this view of the lower mantle had just been challenged [46], and the modern view that the lower mantle has a moderate viscosity and convects quite actively, as discussed in Chapter 4, soon took over. This left the hotspots with no clear reason for being fixed, though they might move more slowly than plates because the viscosity of the lower mantle is greater than that of the upper mantle. Indeed, by now some slow relative motions among hotspots are plausibly resolved [47, 48]. The (approximately) fixed hotspots have nevertheless been very useful in determining plate motions. For our purpose here, which is to examine the science of mantle plumes, it is important to realise that whether plumes are fixed or slow-moving is an issue quite peripheral to the physics of plumes that we are about to discuss.

7.1 Inferring plumes from surface observations

The definition and identification of volcanic hotspots have been debated. Not every intra-plate volcano necessarily qualifies as a hotspot, or at least one to which Morgan's plume hypothesis might relate. Though over 100 have been proposed, around 40 have gained general acceptance, and a representative selection is shown in Figure 7.2.

The strongest and clearest hotspots and hotspot tracks allow some straightforward inferences about the mantle process that gives rise to them, even without the physical theory that we will soon get to. However, there are hotspots and tracks that are more complicated or less clear-cut than the classic case of Hawaii. For example, some of the volcanism in the South Pacific does not form simple linear chains, and a few linear chains do not seem to have a clear age progression. Such examples may well involve additional processes, as will be discussed later. Nevertheless, many hotspots do manifest the relative simplicity of the Hawaiian hotspot.

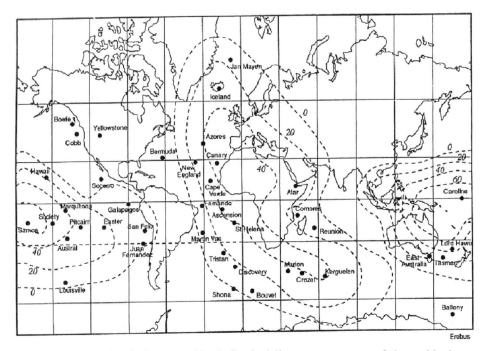

Figure 7.2. Volcanic hotspots (dots). Dashed lines are contours of the residual geoid [49], which is sensitive to long-wavelength gravity variations. From Duncan and Richards [50]. Copyright by the American Geophysical Union.

The occurrence of isolated, persistent and relatively slow-moving volcanism far from plate margins implies an independent and persistent source of magma that is not directly involved with plate motions. The localisation of the volcanism, generally within a radius of a few tens of kilometres, implies that the source is laterally narrow, no more than about 100 km in diameter. The persistence of the volcanism for tens of millions of years (at least 90 Myr in the case of Hawaii) implies a large or self-renewing source. The slow lateral motion implies that the source is below the zone in which the mantle moves with the surface plate, as both Wilson and Morgan inferred. The persistence and slow motion together suggest a vertically extensive source, or in other words a column. The column is around 100 km in diameter but would extend at least 1000 km into the mantle, based on the kinds of convective flow shown in Chapter 6. Thus we have inferred a picture of a tall and remarkably narrow mantle column.

The volcanism of hotspots of course requires melting, and this could be due to either higher temperature or a different composition within the column. If it is due to being hotter, then a persistent supply of heat is available at the base of the mantle, where heat from the core is inferred (on independent evidence) to conduct into the mantle, forming a hot thermal boundary layer. Such a hot thermal boundary layer

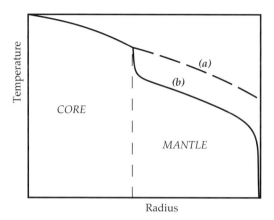

Figure 7.3. Schematic temperature profiles through the Earth. Heat released during the formation of the Earth and the segregation of the metallic core would heat both the core and the mantle (mantle profile (a), dashed). Heat would be lost from the cold surface, and the mantle would cool as a result (profile (b), solid). A thermal boundary layer at the base of the mantle would be formed by heat conducting into it from the core, and this would be a likely place for plumes to be generated.

would form a fluid-dynamically plausible source of our inferred column. Thus we arrive at the essence of Morgan's plume hypothesis.

It is possible to argue that the column has a different composition, but there are difficulties that require *ad hoc* assumptions to overcome. It is not obvious how a compositional source might be continuously renewed, and if it cannot then persistence of the hotspot requires a source of large volume. A problem with this is that a buoyant volume of material will tend to rise towards the surface, and the larger the volume the faster it rises, as we will see a little later. Thus a large buoyant volume is not likely to generate a narrow, persistent compositional plume that could generate hotspot tracks, but rather to rise as a volume and generate a larger, relatively brief eruption, more like the flood basalt eruptions we will also discuss a little later.

The preference for a thermal rather than a compositional column is reinforced by the expectation that the core is very likely to be hotter than the mantle. The gravitational aggregation of the Earth from infalling bodies releases enough heat to vaporise much of the Earth, and even the separation of iron core material from silicate mantle material could heat the Earth by around 2000 °C. Thus the Earth is expected to have formed with a hot interior. It would then begin to cool from the surface downwards, which means the mantle must cool significantly before much heat will conduct from the core. The sequence is illustrated schematically in Figure 7.3. Heat conducting into the base of the cooler mantle from the hot core would then form a hot thermal boundary layer, and such a boundary layer would be a likely source of thermal plumes. Some even argue that the core would start out hotter than the mantle, rather than locally equilibrating with the mantle, as assumed in Figure 7.3. Such a superheated core would strengthen this argument. Estimates of the present temperature of the core, based on extrapolated material properties of iron alloys, commonly require it to be 1000 °C or more hotter than the lower mantle [51], in which case a hot thermal boundary layer would exist at the base of the mantle.

This line of argument implies that thermal plumes in the mantle are virtually inevitable, quite apart from whether they are detectable at the surface. In an Earth-like planet cooling from the surface, heat will conduct from the core, forming a hot thermal boundary layer at the base of the silicate mantle. With sufficient heat flow, the boundary layer will become unstable and generate hot, buoyant upwellings. The favoured form for such upwellings is columns, as we will see shortly. Thus we should expect mantle plumes if there is sufficient heat flow from the core. These would not necessarily be detectable, but volcanic hotspots are a very plausible consequence of such upwellings.

Thus, with few assumptions and no theory, it is possible to infer the presence of hot, narrow columns rising from the base of the mantle under the more prominent volcanic hotspots such as Hawaii. This does not prove the existence of mantle plumes, but it does establish them as a plausible hypothesis with significant evidence in its support. Further arguments will be encountered as we look at experimental and theoretical understanding of buoyant upwellings.

7.2 Hotspot swells, plume flows and eruption rates

As well as the narrow topography of the Hawaiian volcanic chain, a broad swell in the sea floor surrounding the chain is evident in Figure 7.1. This swell is up to about 1 km high and about 1000 km wide. Such a swell might be due to thickened oceanic crust, to a local imbalance of isostasy maintained by the strength of the lithosphere, or to buoyant material under the lithosphere. Seismic reflection profiles show that the oceanic crust is not significantly thicker than normal [52]. Nor can such a broad swell be held up by the flexural strength of the lithosphere [53]. The straightforward conclusion is that the Hawaiian swell is held up by buoyant material under the lithosphere.

Swells like that in Figure 7.1 are evident around many of the identified volcanic hotspots. Other conspicuous examples are at Iceland, which straddles the Mid-Atlantic Ridge, and at Cape Verde, off the west coast of Africa (Figure 2.4). The latter is 2 km high and even broader than the Hawaiian swell, apparently because the African plate is nearly stationary relative to the hotspot [50].

Hotspot swells provide us with important information. They can be used to estimate the rate of flow of buoyancy in the plumes, and from that the flow of

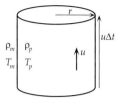

Figure 7.4. Idealised section of a plume, envisaged as a vertical cylinder.

heat can be calculated. The calculation is surprisingly direct, and requires few assumptions about the plume material, other than that its buoyancy is due to heat, rather than to composition.

7.2.1 Buoyancy transported by plumes

Consider material flowing up a plume conduit. Let us envisage the plume as a vertical cylinder with radius r, as in Figure 7.4, so its cross-sectional area is πr^2. Suppose the plume material flows upwards with an average velocity u. Within a short time interval, Δt, the material will flow a distance $u\Delta t$. The volume of fluid that has flowed past a point, say at the bottom of the plume section shown, is then $V = \pi r^2 u \Delta t$. Now, dividing by Δt, the *volumetric flow rate* is $\phi = V/\Delta t$, in other words the volume per unit time flowing past a point:

$$\phi = \pi r^2 u. \tag{7.1}$$

Buoyancy, as we saw in Chapter 5, is the gravitational force due to the density deficit of the buoyant material. The buoyancy of the material in the cylinder section shown in Figure 7.4 is

$$B = g \Delta \rho V,$$

where $\Delta \rho = (\rho_m - \rho_p)$ is the density difference between the plume and the surrounding mantle. This is the buoyancy of the material that has flowed up the plume in the time interval Δt. If we divide B by Δt we can get the *rate*, b, at which buoyancy is flowing up the plume conduit:

$$b = g \Delta \rho \pi r^2 u. \tag{7.2}$$

This establishes the idea of a buoyancy flow rate in an idealised plume conduit. It can be related to hotspot swells, which we will now do.

The way buoyancy flow rate can be inferred from hotspot swells is clearest in the case of Hawaii. The Hawaiian situation is sketched in Figure 7.5, which shows a map view (left) and two cross-sections. As the Pacific plate moves over the rising column of plume material, it is lifted by the plume buoyancy. The weight of the excess topography created by this uplift exerts a downward force, and the buoyancy

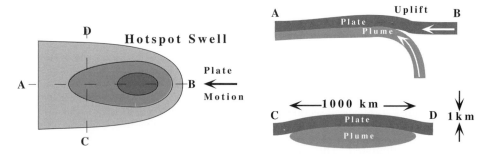

Figure 7.5. Sketch of a hotspot swell like that of Hawaii (Figure 7.1) in map view (left) and two cross-sections (AB, CD, right), showing the relationship of the swell to the plume that is inferred to be below the lithosphere. The swell is inferred to be raised by the buoyancy of the plume material, which arrives as a column (section AB) and then spreads laterally (section CD). This allows the rate of flow of buoyancy and heat in the plume to be estimated.

of the plume material under the plate exerts an upward force. These forces should balance, because the sea floor is not accelerating upwards or downwards and, as discussed above, there is no other force to hold up the swell, such as the flexural strength of the lithosphere.

Since the plate is moving over the plume, the parts of the plate that are already elevated are being carried away from the plume. In order for the swell to persist, new parts of the plate have to be continuously raised as they arrive near the plume. This requires the arrival of new buoyant plume material under the plate (cross-section AB). Thus the rate at which new swell topography is generated will be a measure of the rate at which buoyant plume material arrives under the lithosphere.

The addition to swell topography each year is equivalent to elevating by a height $h = 1$ km a strip of sea floor with a 'width' $w = 1000$ km (the width of the swell) and a 'length' $v\delta t = 100$ mm (the distance travelled by the Pacific plate over the plume in one year at velocity $v = 100$ mm/yr). The weight, W, of one year's worth of new swell is then

$$W = g(\rho_m - \rho_w)wvh. \tag{7.3}$$

The relevant difference in density is that between the mantle (ρ_m) and sea water (ρ_w) because both the sea floor and the moho are raised, and sea water is displaced. (You can work this through by considering the vertical displacement of the moho, and you will find the density of the crust cancels out.)

The requirement that the weight of the topography should balance the buoyancy of the plume material then requires that $b = W$, so the rate of buoyancy flow up the plume is

$$b = g(\rho_m - \rho_w)wvh. \tag{7.4}$$

Using the values quoted above yields $b = 7 \times 10^4$ N/s (newtons per second) for Hawaii.

7.2.2 Heat transported by plumes

If the plume buoyancy is thermal, it can be related to the rate at which heat is transported by the plume, since both depend on the excess temperature, $\Delta T = T_p - T_m$, of the plume (Figure 7.4), as we will now see.

The difference between the plume density, ρ_p, and the mantle density, ρ_m, is due to thermal expansion, which we introduced in Chapter 5. Thus from Eq. (5.3)

$$\rho_p - \rho_m = -\rho_m \alpha \Delta T. \quad (7.5)$$

The rate of flow of thermal buoyancy is then, from Eq. (7.2),

$$b = \pi r^2 u g \rho_m \alpha \Delta T. \quad (7.6)$$

We can relate the rate of heat flow to the volumetric flow rate ϕ (Eq. (7.1)). In Chapter 5 we introduced the specific heat, C_P, which is the amount of heat required to raise the temperature of a unit mass of a material by one degree Celsius. The mass of the cylinder section in Figure 7.4 is $\rho \phi \Delta t$, so its excess heat content, due to its excess temperature, is

$$H = \rho \phi \Delta t C_P \Delta T.$$

The rate at which heat flows up the cylinder is $Q = H/\Delta t$, so

$$Q = \rho_m \phi C_P \Delta T. \quad (7.7)$$

(We can approximate the density in this expression with the mantle density, ρ_m.)

Now if we take the ratio of Q and b, remembering that $\phi = \pi r^2 u$ from Eq. (7.1), then

$$Q = C_P b / g \alpha. \quad (7.8)$$

This is a remarkably simple relationship between Q and b. It involves only g and two material properties, C_P and α. It does *not* depend on the excess temperature of the plume, which cancels out. Nor does it depend on the flow velocity, u, or the plume radius, r.

Thus we can calculate the heat flowing up the Hawaiian plume without having to know ΔT or u or r. Using $C_P = 1000$ J/kg °C and $\alpha = 3 \times 10^{-5}$ °C^{-1}, Eq. (7.8) yields about $Q = 2 \times 10^{11}$ W = 0.2 TW. This is about 0.5% of the global heat flow (41 TW, Table 6.1).

The total rate of heat transport by all known plumes was estimated roughly by Davies [54], and more carefully by Sleep [55], with similar results. Although there are 40 or more identified hotspots, all of them are weaker than Hawaii and many of them are substantially weaker. The total heat flow of plumes is about 2.3×10^{12} W (2.3 TW), which is about 6% of the global heat flow.

Whereas the heat transported by plates accounts for nearly 90% of the heat coming out of the mantle, heat transported by plumes is much less. This suggests that plumes are a secondary form of convection. This conclusion is not surprising, because we estimated the plume heat flow from hotspot swells, and the hotspot swells are a secondary form of topography compared with the mid-ocean rise system (Figure 2.4). We saw in the previous chapter that the mid-ocean rises are intimately related with the plate mode of convection, being due to thermal contraction of plates. In fact, one can relate the heat transported by plates with the negative buoyancy of the plates, due to their thermal contraction, and arrive at the same relationship as Eq. (7.8) [1].

7.2.3 Volume flow rates and eruption rates of plumes

The buoyancy flow rate of a plume was estimated from the swell size without knowing the plume temperature. However, if we do have an estimate of plume temperature, it is possible to estimate the volumetric flow rate of the plume. It is instructive to compare this with the rate of volcanic eruption.

From Eq. (7.6), using Eq. (7.1) for the volumetric flow rate ϕ, we get

$$b = \phi g \rho_m \alpha \Delta T$$

and thus

$$\phi = b / g \rho_m \alpha \Delta T. \tag{7.9}$$

From the petrology of erupted lavas, plumes are estimated to have a peak temperature of 250–300 °C above that of normal mantle [56]. Taking $\Delta T = 300$ °C and using our previous value for Hawaii ($b = 7 \times 10^4$ N/s), with our usual values for the other quantities, this yields $\phi = 240$ m^3/s $= 7.4$ km^3/yr.

It is interesting to compare this with the rate at which the Hawaiian swell is generated. The volume added every year is wvh, and using $h = 1$ km, $w = 1000$ km and $v = 100$ mm/yr, we find that 0.1 km^3 of new swell is raised every year. This is only 1.4% of the volumetric flow rate of the plume, which reflects the fact that the uplift is due to the thermal expansion of the plume material.

We can make a rough estimate of the velocity of flow in the plume, based at this stage only on the inference from the localisation of active volcanism that the plume seems to be 100 km or less in diameter. Equation (7.1) gives us

$$u = \phi / \pi r^2, \tag{7.10}$$

so using $r = 50$ km yields $u = 3 \times 10^{-8}$ m/s $= 0.9$ m/yr. This is about 10 times the velocity of the faster plates. Thus the material in the plumes may flow upwards relatively rapidly, although of course 1 m/yr is still slow by human standards.

It is also interesting to compare the volumetric flow rate with the rate at which magma has been erupted along the Hawaiian volcanic chain. The rate at which the volcanic chain has been constructed over the past 25 Ma has been about 0.03 km^3/yr [57, 58]. This is very much less than the plume volumetric flow rate. It implies that only about 0.4% of the volume of the plume material is erupted as magma at the surface. Even if there is substantially more magma emplaced below the surface, such as at the base of the crust under Hawaii [52, 59], the average melt fraction of the plume is unlikely to be much more than 1%.

Since the magmas show evidence of being derived from perhaps 5–10% partial melting of the source [56, 60], this probably means that about 80–90% of the plume material does not melt at all, and the remainder undergoes about 5–10% partial melting. This result is important for the geochemical interpretation of plume-derived magmas.

7.2.4 Heat flow from the core

If plumes rise from a hot thermal boundary layer formed by heat conducting out of the core into the base of the mantle, then the plume heat flow should be similar to the rate of heat loss from the core. This seems to be true, though estimates of the core heat flux are not easy, and there are some complications in the plume story as well.

The first complication is that the estimate of plume heat flow rate should include the heat carried by plume heads, which will be discussed a little later in this chapter. Hill *et al.* [61] used the frequency of flood basalt eruptions in the geological record of the past 250 Ma to estimate that plume heads carry approximately 50% of the heat carried by plume tails. Thus the total heat flow rate in plumes would be approximately 3.5 TW, still less than 10% of the global heat flow rate.

The second complication is that some careful modelling has shown that the heat carried by plumes is greater in the deep mantle than in the shallow mantle [62–64]. This is because the temperature excess of plumes is larger at depth. There are two contributions to this effect, arising from subtleties of convection and adiabatic gradients, though we don't need to go into details here. Combined, these effects yield a temperature difference between the plume and the surrounding mantle of 400–600 °C at the bottom versus 200–300 °C at the top. Correspondingly, the excess heat transported by the model plumes in the deep mantle is about 2–3 times larger than in the shallow mantle. Thus the plume heat flow is likely to be greater than our estimate above by a factor of 2, with an upper limit of a factor of 3. Thus plumes are inferred to be carrying 2–3 times 3.5 TW, i.e. 7–10 TW.

The heat flow from the core is not only hard to estimate, it is entangled with questions about the thermal evolution of the core and mantle, the rate of growth of the inner core, the generation of the Earth's magnetic field, and the controversial possibility of some radioactivity in the core. These topics are well beyond the scope of this book, so only the main points will be noted here.

Constraints on the heat flow from the core have been deduced from the energy requirements of the geodynamo. There are considerable uncertainties, particularly in the thermal conductivity in the core [65] and in the energy dissipation associated with the dynamo. Lay *et al.* [51] quote a range of 3–8 TW for the heat flow out of the core, but some studies deduce values up to 13 TW [64, 66]. Higher heat flows would imply a relatively young inner core. This in turn would require even higher heat flows in the past, because the dynamo is less efficient without inner core crystallisation, and implausibly high core temperatures are then implied to have existed early in Earth's history. To avoid this difficulty, it has been proposed that the core contains radioactive ^{40}K, which would generate extra heat and prevent the core from cooling so rapidly. However, the geochemical conditions required to sequester potassium into the core ought to have left other clear geochemical signatures that are not observed, so this possibility is doubtful. Details of these arguments, with references, can be found in Davies [65] and Lay *et al.* [51]. Davies [65] presented thermal evolution models of the core in which all constraints were plausibly met and the present heat loss from the core is 5–7 TW. We will encounter a similar evolution model in Chapter 9 and Appendix B.

Thus the estimates of core heat loss and of plume heat flow near the bottom of the mantle are comparable. This supports Morgan's proposal that plumes come from a thermal boundary layer at the base of the mantle.

7.3 The dynamics and form of mantle plumes

Although the preceding discussion is based on observations and fairly direct inferences from the observations, without appeal to theory, in fact there is a well-developed physical theory of thermal plumes. It is based on experiments, numerical models and fairly simple theoretical ideas calibrated by the experiments and models. The result is a good understanding of how thermally buoyant upwellings in the mantle ought to behave. This understanding is good enough to be quantitatively predictive and therefore to be tested. Assertions by some plume sceptics that the plume hypothesis is infinitely adaptable and so not testable and not scientific are therefore not true.

On the other hand, there are observations of non-plate volcanic activity that rather obviously do not fit the predictions of the thermal plume model. Geochemical interpretations of hotspots indicate that plumes do not have the same composition

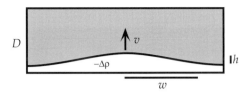

Figure 7.6. Sketch of two layers of fluid, the lower layer with a density lower by $\Delta\rho$. The total depth of the two layers is D. An instability is developing, with a width w and an amplitude h. The highest point of the instability is rising at velocity v. This instability is known to fluid dynamicists as the Rayleigh–Taylor instability.

as ambient mantle, and there are demonstrations that compositional density will considerably complicate the behaviour of plumes, whose buoyancy is affected by both temperature and composition. Because the behaviours are complicated, and have not yet been explored much, it is difficult to say that this accounts for the observed complications, but a plausibility case can be made for some volcanism. There is other localised volcanism that does not seem to have much connection with plumes, so there may indeed be other phenomena occurring in the mantle, but evidently they are not major sources of volcanism or tectonics.

7.3.1 Outline of plume dynamics

I will start by briefly outlining the theory of thermal plumes. Some of the details are covered in following subsections, which you can read or skip, as you please. Our understanding of thermal plumes begins with the instability of a layer of fluid overlain by fluid of greater density, so the lower layer is buoyant (as sketched in Figure 7.6). The buoyant layer will tend to rise, but there is a preferred horizontal scale or spacing of the resulting upwellings. This scale is approximately the total depth, D, of the two layers. Upwellings spaced much closer than D or much further apart than D can grow, but only slowly, and they are overwhelmed by the faster growth of upwellings spaced about D apart.

This preferred spacing of upwellings means that there will not be thousands of tiny upwellings, nor one or two huge upwellings. In other words, it implies that upwellings will tend to be of a certain size. Theoretical estimates and numerical experiments for conditions in the lower mantle indicate that the spacing of upwellings will be of the order of 1000 km, and they will initially form blobs about 400 km in diameter.

Experiments and some theory by Whitehead and Luther in 1975 [67] showed that upwellings tend to take the form of columns rather than sheets. They also showed that, if the upwelling fluid has a lower viscosity than the surrounding fluid, the upwelling takes the form of a roughly spherical 'head' and a relatively narrow

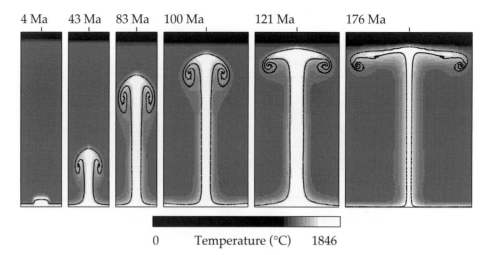

Figure 7.7. Sequence from a numerical model in which a plume grows from a thermal boundary layer. Panels show cross-sections of an axisymmetric plume. The model is scaled to the mantle. A line of passive tracers marks fluid that was initially in the thermal boundary layer. The spiral structure is due to thermal entrainment into the head (see text). The viscosity is temperature dependent, the value at the ambient mantle temperature being 10^{22} Pa s. The bottom boundary is maintained at 430 °C above the interior temperature, and the viscosity of the hottest fluid is about 1% of the ambient viscosity.

'tail'. These results have been confirmed in other experiments and in numerical models. An example of the head-and-tail structure from a numerical model is shown in Figure 7.7.

The head-and-tail structure only occurs when the viscosity of the plume material is less than that of the surrounding material. It comes about as follows. A minimum head size is required before it can detach from the thermal boundary layer and rise through the mantle. The rising head then forces a path through the higher-viscosity surrounding mantle. Once a path is forged, low-viscosity fluid requires only a relatively thin conduit to continue flowing after the head. The lower the viscosity, the thinner the conduit may be. The effect is illustrated quantitatively in Figure 7.8.

The spiral structure evident in Figure 7.7 comes about because of thermal entrainment into the plume head. As the warm head moves up through ambient mantle, it warms adjacent material by thermal conduction, forming a thermal boundary layer around the head, as sketched in Figure 7.9. This makes the newly warmed material buoyant. As the head moves up, its internal circulation wraps some of the thermal boundary layer material into itself, next to the plume tail material. The internal circulation occurs because the fluid flowing up the plume tail is the fastest-rising fluid, and the outer parts of the plume head are resisted by the surrounding fluid,

7.3 The dynamics and form of mantle plumes

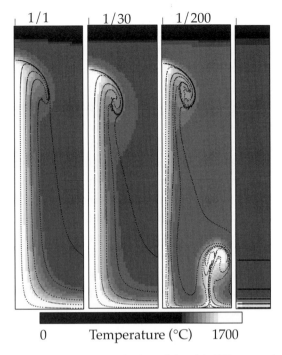

Figure 7.8. Plumes from three numerical models with different ratios of minimum plume viscosity to ambient viscosity, respectively, 1, 1/30 and 1/200. The tail is thinner for lower-viscosity plumes. The models are axisymmetric about the left-hand side of the panels. Several lines of tracers mark fluid initially from different levels near the base of the box. The initial configuration is shown in the right-hand panel. A secondary instability has developed in the 1/200 model.

Figure 7.9. Sketch illustrating thermal entrainment into a hot plume head. The dashed line outlines a thermal boundary layer formed as heat conducts out of the head. As the plume head rises, some of the newly heated material is swept into the head by the internal circulation of the head. The black patches are fluid initially from the thermal boundary layer. Symbols show the head radius, R, a typical thermal boundary layer thickness, δ, the upward velocity of the head, U, and the volumetric flow rate up the tail, ϕ_p.

which is not moving upwards. Thus the fluid on the axis rises through the head and spreads out towards the 'equator' of the head.

A relatively simple theory developed by Griffiths and Campbell [68] predicts that a plume head will start with a diameter of about 400 km near the bottom of the mantle and grow to a diameter of about 1000 km near the top of the mantle. The growth of the head is due both to new fluid flowing up the tail into the head and to surrounding fluid being thermally entrained into the head as illustrated in Figure 7.9. Simple theory, which will be presented below, also shows that a tail diameter of around 100 km (in the upper mantle) allows a sufficient flow up the tail to account for the buoyancies and heat flows inferred in the previous section.

To summarise this section, the hot thermal boundary layer that we expect at the bottom of the mantle is likely to be unstable and to generate buoyant upwellings. These upwellings are more likely to take the form of columns rather than vertical sheets. The upwelling material will have a lower viscosity than the ambient mantle, and this will result in the upwelling forming into a head-and-tail structure. The roughly spherical head will start with a diameter of about 400 km and grow to about 1000 km in diameter near the top of the mantle. The growth is due to thermal entrainment of surrounding material into the plume head, as well as to more material flowing up the plume tail. The tail diameter of about 100 km is sufficient to account for observed plume tail flows.

Simplified versions of the required theoretical results will be given in the following subsections. The discussion so far will have conveyed the essential physics of mantle plumes, but the simple theory shows that plumes are predicted by well-quantified physics. The application of this physical understanding to observations of the Earth is taken up again in Section 7.4.

7.3.2 The Rayleigh–Taylor instability

We can analyse the instability of a buoyant fluid layer using the same approach as we used in Chapter 4 on flow and viscosity and in Chapter 5 on convection.

If the interface between the two layers of fluid in Figure 7.6 were perfectly horizontal, the fluid would not move, even though the lower layer is buoyant. Of course, nothing in nature is perfect, and there will be undulations in the interface, depicted as a simple bulge in the sketch. Because the fluid in the bulge is less dense than the fluid on either side of it, it will generate a buoyancy force, B, proportional to the mass deficit in the bulge. The mass deficit is the density deficit times the volume of the bulge. We will do the usual rough approximating here, so let's take the volume to be simply wh, the width times the height of the bulge, which we will assume to be quite small. Putting these things together, we get

$$B = g\Delta\rho wh. \tag{7.11}$$

The highest point of the interface is depicted as rising at a velocity v (so h is increasing with time). This motion will induce deformation in the fluid, and therefore viscous resistance. In Chapter 4, rate of deformation was measured by a velocity gradient, which is equivalent to a strain rate. A representative velocity gradient here is v/w, assuming the velocity falls away towards zero to the left or right of the bulge. The resulting viscous stress is then

$$\tau = \mu v/w.$$

This stress acts over the width of the bulge, so the total resisting force (per unit length in the third dimension of this cross-section) is

$$R = w\mu v/w = \mu v. \tag{7.12}$$

In the mantle, flow is so slow that acceleration can be totally neglected, as we have discussed previously. This means that there should be no net force acting on the fluid, or in other words the two forces B and R should balance each other. Thus, equating B and R, we get

$$\mu v = g\Delta\rho w h$$

or

$$v = (g\Delta\rho w/\mu)h. \tag{7.13}$$

This says that the velocity is proportional to h, or in other words that the higher the bulge is, the faster it increases its height. This is a familiar equation in basic calculus, and it signifies exponential growth (Appendix A, Section A.3). If the bulge had an initial height h_0, then subsequently its height would be

$$h = h_0 \exp(t/\tau), \tag{7.14}$$

where t is time, exp denotes the exponential function and

$$\tau = \mu/g\Delta\rho w \tag{7.15}$$

has the dimensions of time. As explained in Section A.3, τ is related to a doubling time: $\tau_2 = \ln(2)\tau$. This means that the height h doubles every 0.693τ.

Let's pause to look at what this analysis means so far. It says, via Eq. (7.13), that the higher the bulge, the faster it grows. It says the same thing via Eq. (7.15): h doubles with the passage of every time interval $t = \tau_2$. In other words, once the bulge starts to grow, it undergoes runaway growth: it is unstable. These equations describe an *instability*. Although the buoyant layer moves very slowly at first, if its upper interface is close to horizontal, any undulation with a horizontal scale similar to D will undergo runaway growth, and the layer will eventually break up into a series of rising blobs, much as we have seen in Figure 6.2.

There are two more aspects of the Rayleigh–Taylor instability and its application to a thermal boundary layer that can be analysed. One is that there is an optimum scale (i.e. an optimum value of w) that maximises the growth rate of the bulge. The other is that if the bulge grows too slowly then thermal diffusion can wipe it out, and the instability dies. I will not go through even the simple analyses of the first point here. It can be found in *Dynamic Earth* [1], and more rigorous mathematical treatments can be found in other texts, such as Turcotte and Schubert's *Geodynamics* [53]. Here I will just describe the physics in words.

The growth rate given by Eqs (7.13)–(7.15) is larger for larger w. However, the analysis is only valid for $w < D$, the total layer depth (Figure 7.6). When w is larger than D, the dominant viscous resistance comes from horizontal flow, and the more relevant velocity gradient is w/D. The result of this change is that, for large w, as w gets larger, the growth rate gets smaller, the opposite of the analysis for small w. The growth rate is a maximum when $w \sim D$, and the minimum growth time is approximately

$$\tau_{RT} = \mu / g \Delta \rho D, \tag{7.16}$$

where the subscript RT denotes the Rayleigh–Taylor timescale. This establishes that there is an optimum horizontal length scale or wavelength for which growth of the instability is fastest. This not only gives us an indication of the size of expected upwellings, but also tells us that there will not be flocks of tiny upwellings. The latter idea has sometimes been invoked to argue that there may be many small plumes that elude detection because they are small. However, the idea is fluid-dynamically implausible.

If the lower fluid layer of Figure 7.6 is buoyant because it has a different composition, then there is no more to say about its instability. If, however, it is buoyant because it is hotter, then we can go beyond the basic Rayleigh–Taylor analysis, because there is more physics to consider. Thermal diffusion will tend to smooth out any temperature variations. In Section 5.4 we examined thermal diffusion and found that the time taken for temperature to diffuse over a distance d is proportional to d^2. Equation (5.15a) expresses this as $t = d^2/\kappa$, where κ is the thermal diffusivity. The time for diffusion to act across the whole fluid layer, depth D, will then be

$$\tau_D = D^2/\kappa, \tag{7.17}$$

where the subscript D denotes a thermal diffusion timescale.

We can now think about two physical processes that compete. Bulges in the unstable fluid interface grow, and the timescale of that growth is τ_{RT}. On the other hand, thermal diffusion will tend to smooth out the bulge, with a timescale like τ_D. Which process wins will depend on the ratio of the timescales:

$$\tau_D/\tau_{RT} = g\Delta\rho D^3/\kappa\mu. \tag{7.18}$$

If this looks familiar, it should – it is the Rayleigh number, given by Eq. (5.19). So, we have a new insight into what the Rayleigh number tells us. If τ_D is much smaller than τ_{RT}, then thermal diffusion is rapid compared with the growth of the layer instability, and we might expect that the instability will be inhibited. On the other hand, if τ_D is much larger than τ_{RT}, thermal diffusion will be relatively slow, and the instability might develop with little restraint.

It turns out that there is a threshold value of Ra below which no convection occurs. The threshold or critical value, Ra_c, varies with circumstance, but is typically about 1000. This fact was established by Lord Rayleigh, hence the number bears his name. Convection with Ra just above Ra_c takes the form of simple, regular cells, similar to that shown in Figure 6.2. Such cells had been observed experimentally by Bénard, and this form of convection is known as Rayleigh–Bénard convection. It is the form of convection most commonly shown in textbooks, and Rayleigh's analysis is often the starting point of discussions of convection (e.g. Turcotte and Schubert [53]). However, convection in the Earth's mantle has Ra well above Ra_c and it would not occur in simple cells, even if the plates were not present, as you can see from Figure 6.4.

With insight from Eq. (7.18), we can see that the reason convection does not occur below a threshold value Ra_c is that the initial instability cannot establish itself faster than it is removed by thermal diffusion. Incidentally, we might have expected the critical Rayleigh number to be more like 1 than 1000. The reason it is so large is probably that we have used the time it takes for diffusion to act across the whole fluid layer. However, diffusion may only need to act over a thin thermal boundary layer, and if we used, for example, $D/30$ as our diffusion length scale, then the critical ratio of timescales would be about 1.

7.3.3 Rate of flow up a plume tail

It is useful to consider flow up a plume tail in more detail, as we can get a more physically based estimate of the radius of a plume tail. It will also help us to understand how most plume tails can be relatively thin.

Figure 7.10(a) shows a parcel of fluid within a conduit of buoyant fluid. We will assume that the fluid outside the conduit is not moving, so the velocity of the fluid in the conduit will vary from a maximum in the centre of the conduit to zero at the

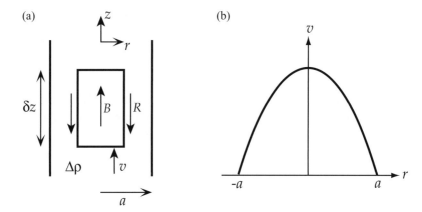

Figure 7.10. (a) A parcel of fluid in a conduit of buoyant fluid. Coordinates are r and z. The conduit has radius a and the fluid parcel has radius r and height δz. The fluid has a density deficit $\Delta \rho$ relative to fluid outside the conduit. The fluid parcel has a buoyancy force B and is resisted by a viscous resistance force R. (b) The variation of velocity across the conduit.

sides. We can look at the balance of the buoyancy force and the viscous resisting force on this parcel. The buoyancy force is

$$B = -g\Delta\rho\pi r^2 \delta z, \qquad (7.19)$$

where the minus sign is because a negative density difference should yield a force in the upward, positive direction. To get the viscous resistance, we need a representative velocity gradient. If the parcel is moving upwards at velocity v, then v/r will be suitable. The resisting viscous stress is then $\mu v/r$. This stress acts over the surface of the parcel, which we take to be a vertical cylinder. Its surface area is its circumference, $2\pi r$, times its height δz. Thus the resisting force is

$$R = 2\pi r \delta z \mu v / r = 2\pi \delta z \mu v. \qquad (7.20)$$

Balancing these forces then yields

$$v = -(g\Delta\rho/2\mu)r^2.$$

Although this analysis is a bit rough, a more rigorous analysis, given in *Dynamic Earth* [1], yields the same result. However, this result is not complete, because it says that the velocity is zero on the axis of the conduit and non-zero and negative at the edge, where, $r = a$. We want a solution in which the velocity is zero at the edge, so we must add a constant velocity (which doesn't change the resisting viscous force) equal to $(g\Delta\rho/2\mu)a^2$, so we get

$$v = (g\Delta\rho/2\mu)(a^2 - r^2).$$

7.3 The dynamics and form of mantle plumes

The rigorous analysis yields a slightly different constant, so finally

$$v = (g\Delta\rho/4\mu)(a^2 - r^2). \tag{7.21}$$

This solution is sketched in Figure 7.10(b). As intuition would suggest, the velocity is a maximum along the axis and falls off smoothly to zero at the sides.

Now let's look at what happens if the radius of the conduit is doubled. The velocity at the axis, where $r = 0$, is $v_a = (g\Delta\rho/4\mu)a^2$, which depends on the square of the radius, so v_a goes up by a factor of 4. However, if the radius doubles, then the area of cross-section of the conduit, which is πa^2, also goes up by a factor of 4. Therefore, the volumetric flow rate – the volume of fluid that flows past a given point per unit time – goes up by a factor of $4 \times 4 = 16$. A rigorous calculation of the volumetric flow rate, ϕ, integrating the flow in concentric circles out from the axis, yields

$$\phi = \left(\frac{\pi g \Delta\rho}{8\mu}\right) a^4. \tag{7.22}$$

Thus doubling a increases ϕ by $2^4 = 16$, as we already deduced.

If we look at this from the other way around, it implies that large changes in the flow rate can be accommodated with only moderate changes in plume radius. The estimates of plume flow rates based on hotspot swells that were discussed in Section 7.2 vary by over a factor of 10 between Hawaii (the strongest) and some of the weaker plumes for which estimates are still possible. Our analysis here implies that the radii of these plumes will not differ by more than about a factor of 2. This implies, for example, that the seismic detectability will not differ greatly among these plumes.

We estimated earlier that the Hawaiian plume has a volumetric flow rate of 240 m^3/s, or 7.4 km^3/yr. We can use Eq. (7.22) and previous values of other quantities to calculate the plume radius. First, assuming the plume is 300 °C hotter than ambient mantle, $\Delta\rho = \rho\alpha\Delta T = 30$ kg/m^3. The viscosity of the upper mantle is not very accurately determined, but a value of 3×10^{20} Pa s is plausible. The plume material would then be as much as 100 times less viscous, so we can assume 3×10^{18} Pa s. Substitution of these values into Eq. (7.22) yields $a = 49\,700$ m or 49.7 km. The agreement with our previous 'guesstimates' of a plume diameter of about 100 km is not to be taken too seriously, because there is considerable uncertainty in some of these numbers. On the other hand, Eq. (7.22) implies that the radius is not too sensitive to these uncertainties. For example, if the viscosity within the plume is 10^{19} Pa s, the calculated radius is only increased by a factor of 1.3 to 65 km. We can therefore conclude with some assurance that plume diameters of about 100 km are fluid-dynamically quite plausible.

Figure 7.11. Sketch of a buoyant plume head of radius r rising with velocity v through a surrounding fluid of viscosity μ. The plume head has a density deficit of $\Delta \rho$ and a viscosity μ_h.

7.3.4 Rate of ascent of a plume head

A blob of buoyant viscous fluid rising slowly through another viscous fluid will assume a spherical shape. This may seem plausible, but it is not so obvious when you realise that a small bubble in water is spherical mainly because of surface tension. Nevertheless, it is observed in experiments. A plume head is also observed to be approximately spherical, though in the thermal plumes of numerical experiments (Figure 7.7) the temperature gradients within the plume head distort it from the spherical. Thus we may approximate a plume head as a buoyant sphere. We can estimate its rate of rise by following the force balance approach that we have already been using. Figure 7.11 is a sketch of a rising plume head.

The buoyancy force of the plume head is

$$B = 4\pi r^3 g \Delta \rho / 3.$$

The rising head drags some of the surrounding fluid up with it as it passes. Fluid adjacent to the head will therefore rise with velocity v, whereas fluid one or two radii away will have a smaller velocity. We may therefore use v/r as a useful measure of the velocity gradients set up in the surrounding fluid by the passage of the head. Therefore viscous shear stresses of the order of $\mu v / r$ will act on the surface of the sphere, generating a resistance force R. The area over which the viscous stresses act is the surface area of the sphere, $4\pi r^2$. Combining these quantities gives

$$R = 4\pi r^2 \mu v / r.$$

Now requiring these forces to balance yields

$$v = g \Delta \rho r^2 / 3\mu. \tag{7.23}$$

Rigorous analyses yield a very similar result, depending on the viscosity within the plume head. A solid or very viscous sphere rises with this velocity, whereas a low-viscosity sphere rises 50% faster. Such differences are not important to our rough estimates. Let us use some values appropriate to mid-mantle depths, where viscosity is about 10^{22} Pa s, density is about 4000 kg/m^3 and thermal expansion is about $2 \times 10^{-5}\,°C^{-1}$. Because of thermal entrainment, the plume head temperature may be only 200 °C higher than ambient mantle. Then $\Delta\rho$ will be 16 kg/m³. A plume head with a radius of 500 km will then rise at 4 cm/yr.

This rise velocity is quite comparable to plate velocities. In the upper mantle it will increase, as the viscosity is lower, but for much of its rise through the mantle this is a plausible velocity. At this rate it would take 50 Myr to rise 2000 km, which would place its highest point near the top of the mantle.

The velocity of flow up the plume conduit that we estimated in Section 7.2.3 is considerably higher than this, about 1 m/yr. This justifies the statements that the plume head will grow as more material flows up the plume tail, as well as from thermal entrainment.

7.3.5 Thermal entrainment into a plume head

The rate of thermal entrainment into a plume head was calculated by Griffiths and Campbell [68], using a thermal boundary layer approach. A summary of that theory is given in *Dynamic Earth* [1] and will not be repeated here, though it is similar in style to the other analyses in this book. The growth of a plume head is due to two things, thermal entrainment of ambient mantle and continued flow of plume material up the plume tail. Taking both of these into account requires a numerical calculation. Results from such calculations by Griffiths and Campbell are given in Figure 7.12.

The thick curves in Figure 7.12(a) are for a viscosity of 10^{22} Pa s, appropriate to the present mid-mantle. They show that the diameter at the top of the mantle is surprisingly insensitive to the plume buoyancy flow. Recalling that the estimated Hawaiian flow is 7×10^4 N/s and that other plumes are weaker, we can see that diameters of about 1000 km at the top of the mantle are predicted. The thin curve in Figure 7.12(a) is for a mantle viscosity of 10^{21} Pa s, which might have applied during the Archaean, and a smaller head is predicted, with a diameter of about 600 km. Figure 7.12(b) shows that the final head diameter is also insensitive to the excess temperature.

7.4 Plume heads and flood basalt eruptions

In 1981, Morgan [69] pointed out that several hotspot tracks emerged from flood basalt provinces. A notable example is the Chagos–Laccadive Ridge running south from the Deccan Traps flood basalt province in western India to Reunion Island in the Indian Ocean (Figures 2.4 and 7.13).

Flood basalts are the largest volcanic eruptions identified in the geological record. They range up to 2000 km across, with accumulated thicknesses of basalt

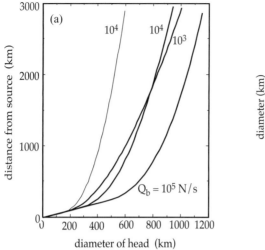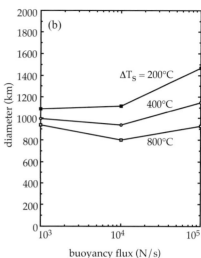

Figure 7.12. (a) Predicted plume head diameter versus height risen in a mantle of viscosity 10^{22} Pa s (thick lines) and 10^{21} Pa s (thin line). Curves are labelled with buoyancy flow rate Q_b. (b) Predicted plume head diameter at the top of the mantle for a mantle viscosity of 10^{22} Pa s and a range of values of the buoyancy flow rate in the plume tail and fluid excess temperature ΔT_s. From Griffiths and Campbell [68]. Copyright by Elsevier Science. Reprinted with permission.

flows up to several kilometres. A map of the main identified flood basalt provinces is shown in Figure 7.13. Total volumes of extrusive eruptions range up to 10 million cubic kilometres, and evidence is accumulating that much of this volume is erupted in less than one million years [70]. It has also been recognised that some oceanic plateaus are oceanic equivalents of continental flood basalts. The largest flood basalt province is the Ontong–Java Plateau, a submarine plateau east of New Guinea.

Morgan proposed that, if flood basalts and hotspot tracks are associated, then the head-and-tail structure of a new plume, which had been demonstrated by Whitehead and Luther [67], would provide an explanation. Figure 7.14 illustrates the concept. The flood basalt eruption would be due to the arrival of the plume head under the lithosphere, and the hotspot track would be formed by the tail following the head. If the overlying plate is moving, then the flood basalt and the underlying head remnant would be carried away, and the hotspot track would emerge from the flood basalt province, and connect it to the currently active volcanic centre, which would be underlain by the active plume tail.

Not a lot of attention was given to Morgan's proposal until Richards, Duncan and Courtillot [71] revived and advocated the idea. Subsequently Griffiths and Campbell [68] demonstrated the thermal entrainment process and argued in more detail for

7.4 Plume heads and flood basalt eruptions

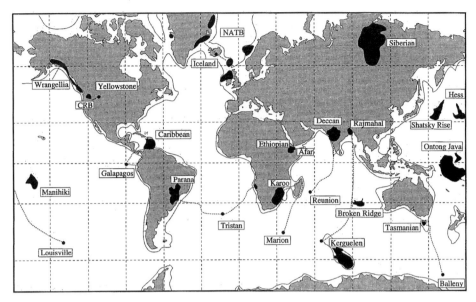

Figure 7.13. Map of continental and oceanic flood basalt provinces. Dotted lines show known or conjectured connections with active volcanic hotspots. After Duncan and Richards [50]. Copyright by the American Geophysical Union.

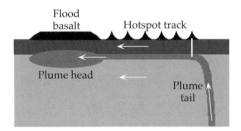

Figure 7.14. Sketch of the way a new plume with a head-and-tail structure can account for the relationship observed between some flood basalts and hotspot tracks, in which the hotspot track emerges from a flood basalt province and connects it to a currently active volcanic hotspot. It is assumed in the sketch that the plate and subjacent mantle are moving to the left relative to the plume source.

the plume head explanation of flood basalts. In particular, Griffiths and Campbell argued that plume heads could reach much larger diameters, 800–1200 km, than had previously been estimated, if they rise from the bottom of the mantle, and also that they would approximately double in horizontal diameter as they flattened and spread below the lithosphere (Figure 7.7). This is in good agreement with the observed total extents of flood basalt provinces, the Karoo flood basalts being scattered over a region about 2500 km in diameter. Campbell and Griffiths [56] then argued that important aspects of the petrology and geochemistry of flood basalts

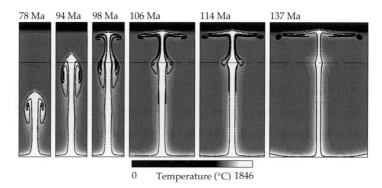

Figure 7.15. Sequence from a numerical plume model including the effect of a mantle in which ambient viscosity increases with depth. The viscosity steps up by a factor of 20 at 700 km depth and has an exponential increase by another factor of 10 through the lower mantle. The plume speeds up as it enters the upper mantle, and narrows as a result. (The effect of a phase transformation is also present in this model, but is not significant here.)

could be explained by the model, in particular the concentration near the centres of provinces of picrites, which are products of higher degrees of melting than basalts. They argued that this can be explained by the temperature distribution of a plume head, which is hottest at the central conduit and cooler to the sides (Figure 7.7).

A potential difficulty with this hypothesis was that the plume head might not produce much melt if it rose under intact lithosphere, because this would keep it below the depth of about 60 km at which the mantle begins to melt [72]. However, Campbell *et al.* [73] pointed out that plumes are expected to melt more readily than normal mantle, because their composition is more basalt-rich [74]. Numerical models [75] showed that this substantially increases the amount of melting, though there was still some deficit. Subsequent more detailed numerical modelling by Leitch and Davies [76, 77] succeeded in accounting for the observed volumes of flood basalts, and in the process found two other factors to be important.

First, the plume head must be grown from a thermal boundary layer. This is because the hottest fluid, coming from the bottom of the thermal boundary layer, flows up the axis and right to the top of the plume head (Figure 7.7).

Second, the vertical viscosity structure of the mantle has an important effect. The upper mantle is 10–100 times less viscous than the lower mantle. This causes the rising plume head to 'neck' down to a much narrower structure as it rises rapidly through the upper mantle (Figure 7.15). It is then easier for the plume to displace material laterally under the lithosphere, so it can penetrate to a shallower depth compared with the plume in Figure 7.7. This considerably increases the amount of melting in the top of the plume head. The result of the combined effects of plume composition, internal thermal structure and rapid ascent through the upper mantle

7.5 Irregular volcanism and thermochemical plumes

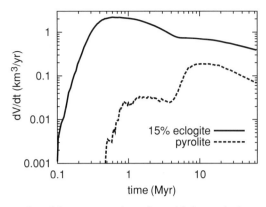

Figure 7.16. Rate of melting versus time from high-resolution computations of melting in plume heads: comparison of melting of normal mantle composition (pyrolite) and a composition including an additional 15% basaltic (eclogite) component. The plumes are grown from a thermal boundary layer and rise through a mantle viscosity structure that decreases in the upper mantle, as in Figure 7.15. From Leitch and Davies [76]. Copyright by the American Geophysical Union.

is that the head can generate millions of cubic kilometres of melt within a timescale of about one million years (Figure 7.16). Thus the plume-head hypothesis seems to account quantitatively for flood basalt eruptions, as well as for their relationship with hotspot tracks.

7.5 Irregular volcanism and thermochemical plumes

Not all intra-plate volcanism fits the simple Hawaiian pattern of an age-progressive line. Two examples from the Pacific basin are shown in Figure 7.17. Nonlinear ridges, splayed sets of ridges, multiple phases of volcanism and no clear age progression are among the features from this region. Some caution is appropriate in drawing conclusions from this, because of possible complications in hotspot tracks. For example, the lack of a clear age progression in a volcanic ridge or chain might reflect overprinting of separate volcanic episodes, jumps of adjacent ridge locations (as is evident with the Ninety East Ridge), complications engendered by continental crust, and slow motion combined with large erupted volumes, as in the case of Iceland. Nevertheless, the complicated morphologies evident in Figure 7.17 make a strong case for something other than a classical thermal plume as a source.

Nor do all hotspot tracks connect with a flood basalt province, as would be expected from the head-and-tail structure of plumes. Clouard and Bonneville [78] argue that, of 14 identified hotspots in the Pacific, only four have plausible or possible connections with an oceanic plateau that might be the product of a plume head.

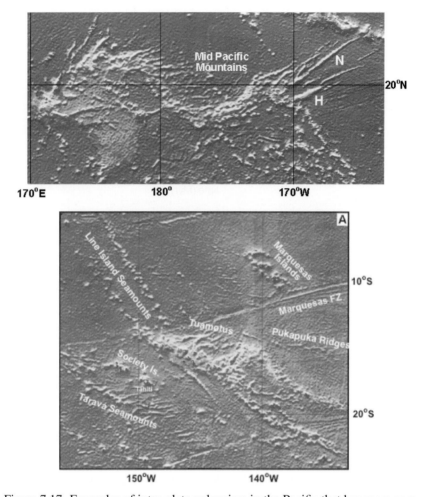

Figure 7.17. Examples of intra-plate volcanism in the Pacific that has more complex morphology and age relationships than a classic hotspot track like Hawaii. The Mid-Pacific Mountains do not form a linear array, and show two zones of splayed ridges. They were erupted at 120–90 Ma ago. The Line Islands show at least two separate volcanic episodes at 68–73 Ma and 81–86 Ma, and the Tuamotus show irregularities and splayed ridges. From Natland and Winterer [79]. Copyright by the Geological Society of America.

They link Easter to the eastern Mid-Pacific Mountains, Louisville to the Ontong–Java Plateau, and, with less confidence, marquesas to Hess Rise and Shatsky Ridge. In addition, the Hawaii hotspot track is interrupted by the Kamchatka subduction zone, and a remnant obducted flood basalt may exist there. Several are indeterminate. Seven do not connect with any obvious ocean plateau, giving rise to the term 'headless plume'.

Some of the observed complications may turn out to be explicable through the effects of compositional buoyancy within plumes. Plumes have long been thought to have a different composition than ambient mantle, and in particular to have a higher proportion of basaltic component, because their source is inferred to contain an accumulation of subducted oceanic crust [60, 74]. Because basalt transforms to an eclogitic assemblage in the upper mantle, and is probably denser than ambient mantle through much of the depth of the mantle, some compositional buoyancy (usually negative) is to be expected [80]. Thus the thermal buoyancy of a plume may be opposed, through much of the plume's ascent, by negative compositional buoyancy. Indeed, there must be a limit to how much basaltic component a plume may contain and still have a net positive buoyancy, and there may be some self-regulation of plumes, only those below the limit being able to rise.

Recent modelling has confirmed that compositional buoyancy may considerably complicate the dynamics of plumes. Lin and van Keken [81, 82] have shown that entrainment of compositionally denser material in axisymmetric numerical plumes can cause plume tail flows to vary erratically and potentially to yield multiple major eruption episodes. Numerical and laboratory studies of three-dimensional thermochemical plumes show plumes that are more irregular in shape and behaviour than the classic thermal plume heads. The laboratory models of Kumagai *et al.* [83] document a spectrum of behaviour. Plume heads with relatively less heavy component may drop some of it as they flatten under the lithosphere. Plumes with more heavy component may stop ascending part way up the mantle, and material with even more may remain in piles near the base of the mantle. In the numerical examples of Farnetani and Samuel [84], much of the plume head stalls under the transition zone but narrow upwellings break through and rise through the upper mantle (Figure 7.18). Such models could provide an explanation for 'headless' plume tracks and a mechanism for small plumes that some have inferred to have arisen within the upper mantle.

These results take plume models into a new realm, with potentially a much richer range of behaviour that may account for a greater proportion of non-plate volcanism. However, the behaviour of thermochemical plumes is clearly complicated and its exploration is still in its early stages, so clear conclusions either admitting or excluding plume sources may not emerge for some time.

7.6 Summary of the plume mode

The existence of volcanic island and seamount chains terminating in isolated, active volcanic hotspots, such as Hawaii, and surrounded by broad topographic swells implies the existence of narrow, long-lived columns of buoyant, rising mantle material. Morgan called these columns *mantle plumes*. The buoyancy and excess

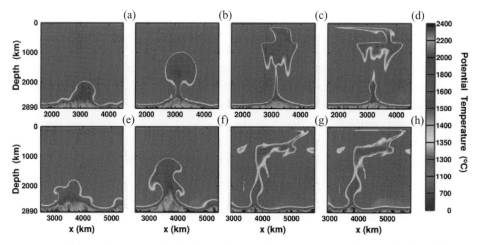

Figure 7.18. Thermochemical plumes from a three-dimensional numerical model [84] in which plume heads stall at the transition zone. Separation of more buoyant material allows small columns to rise through the upper mantle. Copyright by the American Geophysical Union.

melting can be explained if the plumes are 200–300 °C hotter than normal mantle, and their longevity is plausible if they derive from a hot thermal boundary layer. Their higher temperature implies that plumes would have lower viscosity than normal mantle, and fluid dynamics experiments show that the preferred form of low-viscosity buoyant upwellings is columnar, and that new plumes would start with a large, spherical head. Plume heads are calculated to reach diameters of about 1000 km near the top of the mantle, and they provide a plausible explanation for flood basalt eruptions. The association of plume heads with their following plume tails provides an explanation for hotspot tracks that emerge from flood basalt provinces.

Plumes and the flow they drive in surrounding mantle comprise a distinct mode of mantle convection, driven by a hot, lower thermal boundary layer. They are estimated to carry about 3.5 TW of heat through the upper mantle and perhaps 7–12 TW through the lowermost mantle, substantially less than the 30 TW estimated to be carried by plates and much less than the total heat loss from the Earth of about 44 TW. On the basis of this heat flow, and the topography from which it is estimated, plumes thus seem to be distinctly secondary to plates as a mode of mantle convection. Because they are driven by a hot, lower thermal boundary layer, plumes are complementary to the plate mode, which is driven by the cool, top thermal boundary layer. As with the plate mode, the upflow in plumes will be balanced by a passive return flow, downwards in this case.

7.6 Summary of the plume mode

The fact that hotspot locations do not correlate strongly with the current configuration of plates (Figure 7.2; [85]) indicates that the plume and plate modes are not strongly coupled. The implication is that plumes rise through the plate-scale flow without substantially disrupting it. In fact, experiments have shown that plume tails can rise through a horizontal background flow, bending away from the vertical but retaining their narrow tubular form [86]. However, there is a correlation between plume locations, broad geoid highs and slower seismic wave speeds in the deep mantle, indicating that plumes form preferentially away from deeply subducted lithosphere (Figure 7.2, [85, 87]).

Stacey and Loper [88] were apparently the first to appreciate that, if plumes come from a hot, lower thermal boundary layer, then their role is to cool the core, in the sense that they are the agent by which heat from the core is mixed into the mantle. In this interpretation, the role of plumes is primarily to transfer heat from the core *through* the mantle, but not *out of* the mantle. Plates transfer heat *out of* the mantle. Plumes bring heat to the base of the lithosphere, which is mostly quite thick and conducts heat only very slowly to the surface. For example, no excess heat flux has been consistently detected over the Hawaiian swell [89]. While in some cases, like Iceland, the lithosphere is thin and a substantial part of the excess plume heat may be lost to the surface, more commonly much of the plume heat would remain in the mantle, presumably to be mixed into the mantle after the overlying lithosphere subducts.

Thermal plumes do not explain all non-plate magmatism. It is plausible, but not yet clearly demonstrated, that some volcanism that is less regular than the classic age-progressive linear hotspot track may be due to thermochemical plumes, whose behaviour is more complicated and erratic than thermal plumes. It is also plausible that some volcanism has other sources, and several hypothesised mechanisms will be discussed in the next chapter. However, such other mechanisms as may exist evidently account for only a small proportion of the heat transport through and out of the mantle, otherwise their existence would be easier to demonstrate.

8
Perspective

> A fuller picture of mantle convection. Active plates and active plumes. The distinct roles of plates and plumes in heat transport. Plume tectonics cannot replace plate tectonics. How plates and plumes affect each other.
>
> Plumes are not the return mode of plate flow. There is normally no active upwelling under ridges. There is no significant 'decoupling' layer. Return flow is not shallow, 'drag' is small. There is no seafloor 'flattening', though there is some anomalous seafloor elevation. 'Superswells' and residual thermal variations. Layered convection?
>
> Rifts and flood basalts. Superplumes? Small-scale modes? Possible, but evidence is marginal. Edge modes. Drips. Mantle wetspots.

8.1 Separate but interacting

The picture of mantle convection developed so far is of two thermal boundary layers, each driving a distinctive form of convection. Because the two modes of convection are so different, it has been useful to consider the thermal boundary layers separately. Of course the two modes do interact, but not as strongly as in low-Rayleigh-number 'textbook' convection (in which the modes are tightly coupled; Figure 6.2), as we will discuss after a brief assessment of the story so far.

The top thermal boundary layer is very directly implied by all the observations that indicate a steep temperature gradient near the Earth's surface and a shallower gradient further down. The near-surface temperature gradient is directly measured, and the need for a shallower gradient deeper down is implied by the fact that the mantle is not liquid (from seismology) and by temperatures inferred from

8.1 Separate but interacting

magmas reaching the surface. Seismology has also directly resolved the oceanic lithosphere. The bottom thermal boundary layer is inferred from less direct but still robust arguments, principally that the core is likely to be hotter than the mantle.

The two thermal boundary layers behave very differently, and the reason is to be found in their mechanical properties. The rheology of mantle material depends strongly on temperature. At higher temperatures, the main effect is that the viscosity decreases as temperature increases further. This means that upwellings from the bottom thermal boundary layer take the form of columns (rather than sheets), and that the columns start with a large head, with a thinner tail following. On the other hand, at lower temperatures, mantle rheology changes its character, from a deforming (approximately viscous) fluid through an intermediate range to effectively a brittle solid at the surface. This means that the colder parts of the top thermal boundary layer are strong, and move as 'rigid' units. The moving parts, the plates, are separated by narrow faults or shear zones. If a plate descends into the mantle, it usually retains its sheet structure at least within the upper mantle, according to images from seismic tomography [90].

The plates, being pieces of the top thermal boundary layer, are the active components in the plate mode of convection. Because they are strong and move as a unit, the flow under them will be coherent. They also control where upwellings and downwellings occur. The result is a mode of convection in which each plate drives a 'cell' or roll-like flow.

The plumes are also active components, but deriving from the bottom thermal boundary layer and driving a quite different form of flow. The plumes are narrow, columnar upwellings rather than sheets. By conservation of mass there will be a corresponding broad, slow downwelling driven by each plume. Plumes seem to be capable of penetrating the rolls of the plate mode, judging by the low correlation between volcanic hotspots and plate boundaries.

Plates and plumes must be regarded as separate agents playing different roles in the mantle, though the flows they generate do interact, as we will discuss shortly. A fundamental difference is that plates cool the mantle whereas plumes cool the core. We have seen that the cycle of upwelling under a mid-ocean ridge, cooling to form a plate, subduction and reheating in the mantle acts to remove heat from the deep interior of the mantle. The upper thermal boundary layer is where heat is lost from the mantle by conduction. On the other hand, the lower thermal boundary layer is where heat enters the mantle from the core. Plumes rising from this thermal boundary layer transport that heat through the mantle, and most of it stays under the lithosphere and is stirred back into the mantle interior. Only a minor amount conducts through the lithosphere to the surface.

A couple of implications follow immediately from recognising plates and plumes as independent agents playing different roles. One is that 'plume tectonics' is not an alternative to plate tectonics, as has sometimes been suggested for Venus or for the early Earth. If plate tectonics did not operate, the mantle would have to dispose of heat through the top thermal boundary layer in some other way. If the lithosphere were intact, forming a 'one-plate' planet, the heat would have to conduct through the lithosphere. If the top thermal boundary layer adopted some other form of motion – for example, if the lithosphere were too weak to form rigid plates – then that motion would accomplish the heat removal. The activity of plumes is determined by the temperature difference across the bottom thermal boundary layer, and that would not be much affected by a change in the behaviour of the top thermal boundary layer. Thus plumes would continue their activity regardless of how the top thermal boundary layer was behaving.

The second implication of the independence of plates and plumes is that plumes are not the return flow of plates. If there were no heat coming from the core there would be no plumes, but plate tectonics could operate as usual. The return flow of the plates is a broad upwelling between subduction zones that focuses into the mid-ocean ridges at the surface (see Figure 6.3). The return flow of plumes is a broad and very slow downwelling between plumes. Both return flows are passive. The idea that plumes are the return flow of plates probably comes from textbook examples of convection, in which the flows driven by the two thermal boundary layers are tightly coupled, so that the active upwellings superimpose on the return flow from the active downwellings, and vice versa, as can be seen in the last panel of Figure 6.2.

Plates do seem to influence the plume mode, even though many plumes rise under the interiors of plates. The main evidence for this is the correlation of hotspots with geoid highs (Figure 7.2). The geoid reflects deep variations in the density of the mantle, and geoid lows correlate with the locations of subduction during the past 100 Myr or so, plausibly because of accumulations of old subducted lithosphere in the deep mantle [87]. By implication, regions between subduction zones, under the geoid highs, are regions where the deep mantle is flowing upwards in the broad return flow from plate subduction. Plumes occur preferentially in these mantle upwelling regions.

This correlation is understandable, because instabilities in the bottom thermal boundary layer are most likely to occur where it is being thickened by converging deep flow driven by subduction zones (see Figure 8.1). Equation (7.16) shows that the growth of the Rayleigh–Taylor instability in the thermal boundary layer is faster (smaller τ_{RT}) if the layer is thicker.

Plumes may also affect the plate mode of convection. Morgan originally pointed out that the beginning of the opening of the Atlantic seemed to correlate with the start of new plumes [44], and he actually argued that plume tails were a major source of plate driving force. Our analysis of slab and plume buoyancy forces does not support this possibility. However, some cases of rifting do correlate plausibly

Figure 8.1. Sketch of the generation of plumes in the broad zone of upward mantle return flow between subduction zones. Dark grey: hot thermal boundary layer and plumes; light grey: warm zone. The deep converging flow thickens the bottom thermal boundary layer, which increases its instability and its propensity to generate plumes. Such a sequence might account for the burst of plume formation in the Pacific region during the Cretaceous. After Davies and Pribac [91]. Copyright by the American Geophysical Union.

with the arrival of plume heads, sometimes with a time delay [92]. The delays may be due to the lithosphere needing to be weakened by heat conducting up from the plume head. Examples are the jump of rifting from the west to the east of Greenland, the openings of the central Atlantic and the South Atlantic, and rifting of India's west coast. Not all plume head arrivals resulted in rifting, as the examples of the Columbia River Basalts and the Siberian flood basalts testify.

Thus the main effect of plumes on the plate mode seems to be to occasionally trigger the break-up of a plate. The mechanism is probably the uplift by several hundred metres caused by the plume head [93], which would generate a gravity sliding force.

8.2 Common misconceptions

Many ideas were proposed in the early days of plate tectonics in attempts to understand this unfamiliar phenomenon. A few of those ideas have survived and form the basis of the previous chapters. Other ideas have been superseded or shown to be incorrect. However, some of the latter ideas still persist, presumably because the modern understanding of mantle convection is not as widely known as it might be. Some of the more common misconceptions will be noted here.

8.2.1 Plumes are not the return mode of plate flow

We noted this in the previous section. Plumes are active components of convection driven by a bottom thermal boundary layer. They could exist even if there were no plate tectonics.

8.2.2 There is normally no active upwelling under mid-ocean ridges

In the account given in Chapter 6, the plate mode of convection is driven by the negative buoyancy of subducted lithosphere. The idea of an additional, positive

108   *Perspective*

Figure 8.2. Topography of the sea floor near the Eltanin fracture zone in the southeast Pacific (compare with the global map, Figure 2.4). The East Pacific Rise is cleanly offset by the Eltanin transform fault, with no indication of rise topography extending across the fault. This is explained if the rise topography is due entirely to near-surface cooling with no contribution from a putative buoyant upwelling under the rise.

source of buoyancy under ridges is a hangover from simple textbook examples of convection (such as Figure 6.2), which are driven equally by hot and cold thermal boundary layers acting in a coordinated way. We noted in Chapter 3 that the idea actually impeded the understanding of plate tectonics. The point being made here does not apply to mid-ocean ridge segments that have a mantle plume rising under them, such as Iceland, but those are clearly the exception.

Not only is there no need to invoke buoyant upwellings under mid-ocean ridges, but also there is clear evidence against such upwellings. If such active upwellings existed, they would contribute to seafloor topography by lifting ridges even higher than they are. Subsidence of sea floor as it aged would then be due to two causes: thermal contraction, as we discussed in Chapter 6, and moving off the active upwelling. The result would be that sea floor would for a time subside faster than predicted by the square root of age cooling. This would destroy the correlation between seafloor depth and the square root of age shown in Figure 2.5.

The absence of active upwellings is manifest in another way. If there were buoyant upwellings under ridges, the upwellings would have to be offset at transform faults. The fluid upwelling would cause a more continuous uplift than the sharp, faulted offsets of the brittle lithosphere. We would therefore expect some uplift to extend beyond the end of each ridge segment and some uplift of the transform faults connecting spreading centre segments.

Figure 8.2 shows topography of the sea floor near one of the longest transform offsets of any spreading centre, at the Eltanin fracture zone in the southeast Pacific. It is striking how cleanly the whole mid-ocean ridge structure is terminated and offset by the transform fault. Where the spreading centre is terminated by the transform fault, the sea floor on the other side is at a depth normal for its age. There

8.2 Common misconceptions

is no hint of a bulge, due to putative buoyant upwelling under the spreading centre, extending across the transform fault. This topography is difficult to reconcile with a buoyant upwelling from depth, but is readily explicable if the mid-ocean ridge topography is due to the near-surface and local process of conductive cooling, thickening and thermal contraction of the thermal boundary layer (that is, of the lithosphere).

If upwelling under normal mid-ocean rises is passive, then there is no problem with spreading centres that move relative to other parts of the system: they merely pull up whatever mantle is beneath them as they move around the Earth. This solves the puzzle that led Heezen (Chapter 3) to postulate an expanding Earth in order to try to explain how spreading centres could exist simultaneously on both sides of Africa.

8.2.3 There is no significant 'decoupling' layer

There is a zone of low seismic velocity in the upper mantle due to the close approach of the geotherm to the melting temperature [94]. It is plausible that mantle viscosity is also a minimum in this zone. This led to the idea that the zone would 'lubricate' the plates, decoupling them from the deeper mantle and allowing them to move independently. However, a viscosity 4–5 orders of magnitude lower than ambient mantle would be necessary for such decoupling, and there is no expectation of or evidence for such a drastic drop in viscosity. The idea of lubrication also gained support from efforts to model forces acting on plates [95], which concluded that there was little 'drag' acting on the base of plates. However, this can be explained by the mantle under plates moving essentially with the plates, both driven by the sinking lithosphere. We were able, in Chapter 5, to explain the velocities of plates without having to appeal to any decoupling from the underlying mantle (Figures 5.2 and 6.3).

8.2.4 Return flow is not shallow

There was an early assumption that mantle flow implied by plate tectonics was confined to the upper mantle, because the lower mantle had been believed to be so stiff that it was essentially static, as we noted in Chapter 7. Some versions even assumed that the 'return flow' is confined to the presumed low-viscosity zone, only about 100 km thick, under the plates. This idea is closely related to the decoupling layer idea, and the arguments against it are the same. Numerical models show that the return flow of plates penetrates into the lower mantle, even though the viscosity of the upper mantle is a factor of about 30 less than that of the lower mantle, as will be seen in Chapter 9.

110   *Perspective*

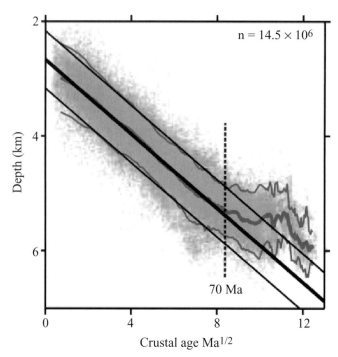

Figure 8.3. Example of globally aggregated depth–age data. Straight lines are a model subsidence ($z = 2648 + 336\sqrt{t}) \pm 500$ m. Curves are the mean $\pm$ two standard deviations from 1 Myr bins. From Hillier [97]. Copyright by the American Geophysical Union.

8.2.5 There is no seafloor 'flattening'

This is a very persistent idea, but there is clear evidence against it. The evidence was presented by Marty and Cazenave [10] in 1989, Figure 2.5 being a selection of their results. Their results show that there are regional variations in seafloor subsidence rates but no persistent tendency to asymptotically approach a constant depth. The alleged evidence for such asymptotic flattening comes from globally aggregating seafloor depths into a single plot. A recent example of an aggregated data set is shown in Figure 8.3. As can be seen, it does not even show the asymptotic flattening it is purported to document. Nevertheless many discussions of the flattening interpretation have continued (e.g. [96–98]) even though the plots of Marty and Cazenave show that the apparent flattening is an artefact of the aggregation.

An accurate characterisation of seafloor depth is that there is an underlying tendency to follow a square root of age subsidence curve (Section 6.4), but there are many regional deviations of older sea floor to higher elevations. The distinction between these characterisations (regional anomalies versus global flattening) is important because it implies quite different physics, as was spelt out by Davies and

8.2 Common misconceptions

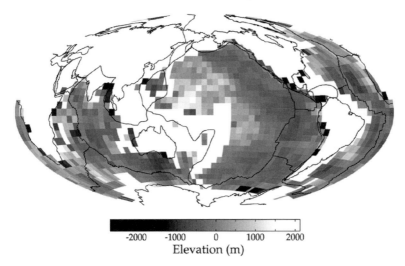

Figure 8.4. Anomalous seafloor topography. An age correction (with a subsidence rate of 320 m/Ma$^{1/2}$) and a correction for crustal thickness using the 5° × 5° grid of Mooney et al. [100] have been removed. Figure courtesy of S. V. Panasyuk, Harvard University [101].

Pribac [91] in 1993. The idea of asymptotic flattening arose from early, limited depth and heat flow data that seemed to show a levelling-off for sea floor older than 70 Ma [99], and it was interpreted in terms of oceanic plates that approach a limiting thickness of about 100 km. If, however, the old sea floor merely shows regional positive depth anomalies, the implication is that plates normally continue to thicken with age, and that some deeper source of buoyancy slows or reverses their subsidence in some regions.

Davies and Pribac pointed out that there are long-wavelength variations in the depths of mid-ocean ridge crests by hundreds of metres, with a maximum of about a kilometre. These can be seen in Figure 2.4; particularly notable is the low elevation of the ridge between Australia and Antarctica, which is about 1 km deeper than normal. These obviously are not due to variations in lithosphere age (it is all age zero), nor are they due to variations in crustal thickness. The only possibility is that they are due to long-wavelength variations in the mantle, plausibly of mantle temperature. A few tens of degrees variation for some hundreds of kilometres depth is sufficient to account for the depth variations.

Some such variations are expected just because the temperature variations due to lithosphere subduction take a long time to be homogenised. If such variations exist under ridge crests, they should be expected elsewhere as well. Therefore, we should expect long-wavelength deviations from square root of age subsidence with amplitudes of hundreds of metres. Just such deviations exist, as is illustrated in Figure 8.4.

There are other plausible sources of low-amplitude topography. Plumes are an obvious source, but they can usually be identified fairly clearly. There are also deep thermochemical 'superpiles' in the lower mantle, one under Africa and one under the central Pacific [102–104]. Simmons *et al.* [104] infer that the relatively high elevation of Africa is due to the pile under Africa.

There is a broad positive anomaly stretching from the central Pacific to the Japan Trench. This anomaly has contributed to the impression of flattening, but its regional character can be seen in Figure 8.4. It was called the Pacific Superswell by Davies and Pribac [91] and interpreted as a residue of the Mesozoic Darwin Rise inferred by Menard [105]. The mantle structure envisaged by Davies and Pribac is illustrated by Figure 8.1.

8.3 Layered convection?

Perhaps the most persistent and sometimes heated debate about mantle convection has been whether the mantle convects in two separate layers or as a single, whole-mantle layer. The initial conception was that the upper mantle, above 660 km depth, convects separately from the lower mantle [106, 107]. As evidence against this picture emerged, a deeper layer, within about 1000 km of the core, was proposed [108], but there are also strong arguments against such a layer. The only layer for which there is clear evidence is the so-called D'' zone within the lowest 200–300 km of the mantle [109, 110].

The original notion of layered mantle convection grew out of the idea, mentioned earlier, that the lower mantle is so viscous as to be immobile, so in the late 1960s convection was presumed to be confined to the upper mantle [111]. When the lower mantle turned out to have a moderate viscosity [46], as presented in Chapter 4, it was presumed by the mid-1970s to convect separately [106]. There were no strong reasons for this presumption [112], but one suggestive reason was that earthquakes in the Wadati–Benioff deep earthquake zones under ocean trenches are confined to the upper mantle. This was interpreted to mean that the plate-related flow was also confined to the upper mantle. In fact, most of the deep zones give no indication of deflecting horizontally and are consistent with flow continuing into the lower mantle [1]. The cessation of earthquakes can be plausibly attributed to the phase transformations that occur at this depth. The geophysical case for whole-mantle convection was being argued by 1977 [112, 113].

When, in 1976, some of the first measurements of neodymium isotopes were interpreted to indicate a primitive zone in the mantle [114], the layered picture was invoked, the lower mantle being presumed to be geochemically primitive, and chemically isolated from the upper mantle [107]. Soon it became evident that

8.3 Layered convection?

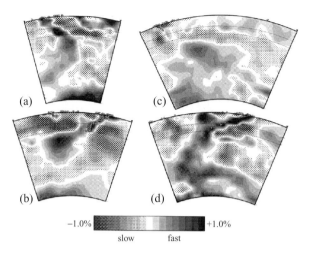

Figure 8.5. Profiles through seismic tomography models of subduction zones: (a) Aegean, (b) Tonga, (c) Central America and (d) Japan. The shading shows variations in the shear wave velocity. Small circles show earthquake locations. From the model S_SKS120 of Widiyantoro [116].

the few data with a putatively primitive signature were probably contaminated by continental crust, but the idea of a primitive lower mantle rapidly became entrenched, in spite of the slender evidence supporting it, and in the face of clear and well-known contrary evidence from lead isotopes [115]. The geochemical issues will be taken up again in Chapter 10.

There are two strong arguments against a separation of convection at 660 km depth, one well known, the other less well known but relevant to subsequent versions of layering. The well-known evidence comes from seismic tomography, and shows zones of high seismic velocity reaching from surface subduction zones well down into the lower mantle (Figure 8.5). These zones are readily interpreted as subducted lithosphere penetrating through the 660 km level and deep into the lower mantle.

The mass flux associated with subducted lithosphere penetrating into the lower mantle rules out any long-term compositional difference between the upper and lower mantles. Lithosphere subducts at a rate of about 3 km^2/yr, has a density of about 3300 kg/m^3 and is about 100 km thick. This amounts to a mass flow of 10^{15} kg/yr. The lower mantle contains about two-thirds of the mass of the mantle, so its mass is about 2.6×10^{24} kg. This means that it takes about 2.6 Gyr to replace the lower-mantle material, at present rates of subduction, and the rates are likely to have been greater in the past, as we will see in Chapter 9. This implies that the lower mantle cannot be chemically isolated from the upper mantle. Some have proposed

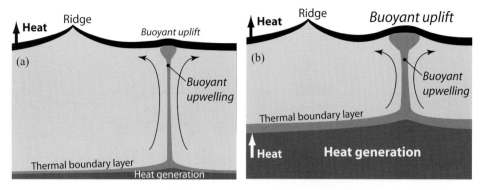

Figure 8.6. Sketches of the relationship between heat flow from depth, buoyant upwellings and topography. (a) The situation inferred in Chapters 6 and 7, in which the only heat from depth is heat from the core plus a small amount generated by radioactivity in the thin D″ layer. (b) The situation proposed by Kellogg *et al.* [108] in which a thick, deep layer contains about half of the Earth's radioactive heat sources. This layer should generate strong upwellings and major topography. After Davies [118]. Copyright by the American Geophysical Union.

that mantle layering existed until relatively recently (perhaps only through the Phanerozoic), but a breakdown of layering ought to have produced dramatic events at the Earth's surface, as we will see in Chapter 9, and there is no evidence for such an event.

The evidence from seismic tomography [90] persuaded most geochemists by about 1997 that there is no barrier to flow at 660 km depth. However, it wasn't long (1999) before a deeper interface was proposed, around 2000 km depth [108], supported by geophysical evidence [117] that is circumstantial at best. The second strong argument against layering also applies to this proposal. The argument is that any such interface should generate strong mantle plumes, and there is no evidence for such plumes in the surface topography of the Earth. The argument is summarised in Figure 8.6.

The picture developed so far in this book is of a strong plate mode of convection and a secondary plume mode (Figure 8.6(a)). About 7–10 TW of heat is carried from the deep mantle by plumes. Most of this heat comes from the core, though perhaps 1 TW might be generated in the thin D″ layer, as we will see in Chapter 10. A heat budget of this model is depicted in Figure 8.7(a) [118]. By the time the plumes reach the shallow mantle, they account for only about 3.5 TW, much less than the ∼30 TW carried by plates. The topography generated by plumes, the hotspot swells, is correspondingly secondary compared with the mid-ocean ridge system, as we have seen.

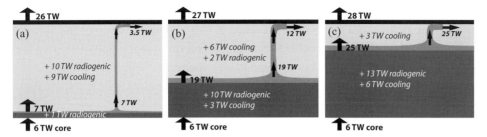

Figure 8.7. Heat budgets of three mantle models [118]. (a) The whole-mantle model, with a thin D″ layer at the bottom. (b) The deep layer model, containing about half the heat-generating elements. (c) The old primitive lower-mantle model, containing about two-thirds of the heat-generating elements. After Davies [118]. Copyright by the American Geophysical Union.

On the other hand, the main reason for proposing the thick layer depicted in Figure 8.6(b) was to accommodate extra trace elements [108]. Meteorite studies provide estimates of the Earth's complement of many refractory trace elements, such as uranium and thorium. Many of these elements tend to be concentrated in the continental crust, but the continental crust accounts for only about half of the total. A major rationale for the earlier primitive lower-mantle model was that it accounted for the balance of these trace elements [107]. When that model became untenable, there seemed to be nowhere these elements could be accommodated, so the deeper layer was proposed as a place that could accommodate them without obviously contradicting geophysical evidence. However, this model does contradict important evidence.

If the deep layer contains about half of the Earth's U, Th and K, the main heat sources, then a large fraction, one-third to one-half, of the Earth's radiogenic heat budget would be generated within it. A reasonable estimate is that about 13 TW would be added to the core heat flow (6 TW; Figure 8.7(b)). This heat would conduct through the top interface of the deep layer, forming a thermal boundary layer that would generate plumes. By the time the plumes rose the 2000 km to the shallow mantle, their heat flow would have attenuated to about 12 TW, much greater than the 3.5 TW inferred from the hotspot swells. In other words, this model implies that hotspot swells should be 3–4 times more prominent than they are (Figure 8.6(b)). They should be quite obvious in Figure 2.4, instead of being rather subtle. The hotspot topography should be nearly half the size of the mid-ocean ridge topography, which is associated with the transport of about 30 TW.

This argument applies to any deep layer that contains a significant proportion of the Earth's radiogenic heat sources. It applies to the old primitive lower-mantle model, which would have contained about two-thirds of the heat sources. The heat budget for this model is shown in Figure 8.7(c). By this estimate, the hotspot topography should be comparable to the mid-ocean ridge topography in extent and volume. Clearly it is not. This argument was made in 1988 [54], though it did not seem to attract much notice, perhaps because it requires a little explanation, whereas tomographic images require little explanation.

Thus the topography of the sea floor not only tells us about the heat transported by the two main modes of mantle convection, but also implies a strong constraint on the structure of mantle convection. The relative smallness of hotspot swells, and the absence of any other topography that can be interpreted as due to buoyant mantle upwellings, implies that only a modest amount of heat can come from below the source of mantle plumes. Because plumes would be generated above any interface within the mantle, not much heat can come from below that interface. Thus any deep layer containing a substantial fraction of Earth's radioactive heat sources is precluded. This conclusion creates a major difficulty in the conventional view of mantle geochemistry. The potential resolution of that difficulty will be taken up in Chapter 10.

8.4 Other modes and causes

Although a fairly clear picture has been painted by the foregoing arguments, there are of course still debates about the mantle convection system. These tend to focus on specifics and details. For example, our knowledge of plate motions for the past 100 Myr or so allows direct modelling of the pattern of mantle flow [119], with application to rather specific questions. This book is intended to convey a basic general understanding, so such topics are not pursued here. However, some issues have played large roles in past debates, and some point to significant lessons, so a few will be reviewed here.

8.4.1 Rifting model of flood basalts

White and McKenzie [72] proposed a theory for the formation both of very thick sequences of volcanic flows found along some continental margins and of flood basalt eruptions. The theory can usefully be separated into three parts. The first part is that the marginal volcanic provinces are produced when rifting occurs over a region of mantle that is hotter than normal because it is derived from a plume. This seems to give a very viable account of such provinces, and it has been quantified successfully using a plume head model [77]. The second part is that all flood basalts can be explained by this mechanism. The third part is that the plume material is derived mainly from a plume tail, since they assumed that plumes are part of an upper-mantle convection system and that plumes therefore derive from no deeper than 670 km. In this case the plume heads would have diameters of no more than about 300 km and volumes less than about 5% of a plume head from the bottom of the mantle [68].

The second part of White and McKenzie's model encounters the difficulty that a number of flood basalt provinces are said, on the basis of field evidence, to have erupted mainly before substantial rifting occurred (e.g. Deccan Traps) or in the absence of any substantial rifting (e.g. Siberian Traps, Columbia River Basalts) [120]. It also fails to explain the very short timescale of flood basalt eruptions, less than 1 Ma in the best-constrained cases. The third part of their model implies that a sufficient volume of warm mantle would take about 50 Ma to accumulate, but, at the time the Deccan Traps erupted, India was moving north at about 180 mm/yr (180 km/Myr) so it would have traversed the extent of the flood basalts in only about 10 Ma. It is implausible that sufficient warm mantle could accumulate from a plume tail under such a fast-moving plate.

The latter two difficulties are avoided by the plume head model of flood basalts, since the flow rate of the plume head is much greater than that of the tail and much of the melting is inferred to occur from beneath the intact lithosphere upon arrival of the plume head, as we saw in Chapter 7. On the other hand, White and McKenzie's discussion of rift-margin volcanism is quite reasonable, and also compatible with the plume head model.

8.4.2 Superplumes

The term *superplume* has been used in many contexts, usually without any clear explanation of what the term might refer to, other than something larger than or different from a 'normal' plume. The question is whether there might be any phenomenon for which the term might be appropriate. There may be one such possibility, but the other uses of the term seem to be either unjustified or so ill-defined as not to be useful.

The term seems first to have been used by Larson [121]. Larson argued that there had been a burst of volcanism in the Pacific during the Cretaceous, and that it correlated with a number of other phenomena of the period. There were indeed several large hotspots active during this time, and possibly other less localised volcanism as well. Larson proposed that the period might be explained by a 'superplume', but offered no definition or physical basis for the term. Because the phenomena persisted over tens of millions of years and at several loci, it seems unlikely that a single mantle upwelling would be responsible. A slightly unusual burst of plume activity seems quite adequate to account for the observations, and Davies and

Pribac [91] later offered a mechanism for such a burst of plumes, as discussed in Section 8.2.5 and illustrated in Figure 8.1.

The term *superplume* quickly gained currency, but not a physical or observational basis. It is of course not difficult to adjust the parameters of a numerical model to produce a very strong upwelling, but that may or may not be consistent with observations, a question that was not given much attention.

As the deep seismic anomalies under Africa and the Pacific began to be resolved [103], the term *superplume* was applied to them. However, they are very broad, narrowing towards the top, and bear no obvious resemblance to either a plume tail or a plume head. Because such large volumes would rise very rapidly if they were purely thermal, it seemed that they must be compositionally heavier, an inference later quantified by Simmons *et al.* [104]. In that case they can plausibly be identified with the piles of denser material that develop in numerical models that include subducted oceanic crust [122, 123], which is denser than normal mantle and tends to settle to the base of the mantle. We will encounter these models in Chapter 10. They suggest that the anomalies are connected with the D″ seismic layer in the lowest 200–300 km of the mantle, and that both are due to the accumulation of old subducted oceanic crust. Where deep-mantle flow converges and turns upwards, the accumulation is swept into piles, which can quantitatively be identified with the deep seismic anomalies [119]. Such piles might be called *thermochemical piles* or *superpiles*, but not plumes.

There is one case in which the much-abused term *superplume* might apply. Simmons *et al.* [104] have imaged the sub-African structure and also separated thermal and compositional contributions to the seismic anomaly. Their results (Figure 8.8) show a structure that is rounded at the top and thinner part way down. This they plausibly interpret as having a positive net buoyancy and an upward velocity, though it might not continue to ascend all the way to the top of the mantle. Because this structure actually has some resemblance to the head-and-tail structure of a thermal plume, it is not unreasonable to call it a superplume.

8.4.3 Small-scale convection

The idea that the lithosphere approaches a constant thickness contributed to the development of the idea of a pervasive mode of 'small-scale' convection confined to the upper mantle. The constant-thickness lithosphere model implied a heat input to the base of the lithosphere, in order to maintain an asymptotic steady-state heat flux at the surface, and it was proposed that this was due to some form of sub-lithospheric small-scale convection. Initially this was supposed to be convection cells of the scale of the upper mantle, to which this mode of convection was assumed to be confined [124–126].

8.4 Other modes and causes 119

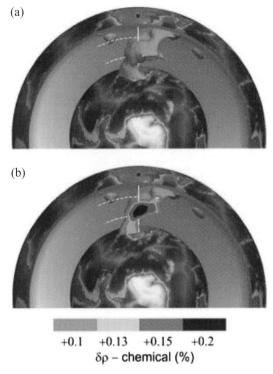

Figure 8.8. Visualisation of the thermochemical structure under Africa. (a) View exhibiting a plume-like morphology with a large-volume rounded structure centred at 1800 km depth directly beneath the southern African continent. (b) View of the interior of the structure where a positive chemical density field is inferred. After Simmons et al. [104]. Copyright by the American Geophysical Union.

A later variation was that small-scale convection is driven by instability of the lowest, softest part of the lithosphere [127]. However, it was assumed rather than demonstrated that the lower lithosphere had the requisite mobility, the thermal boundary layer being assumed in that study to have a stepped viscosity structure. Subsequent evaluation using a more appropriate temperature-dependent viscosity [128] showed that it is far from clear that such convection would have significant amplitude, or even occur at all.

The proposal for an upper-mantle scale of convection, that is for cells of about 650 km depth and a comparable width, encounters the topographic constraint already discussed for any form of upper-mantle convection: there ought to be substantial topographic signatures of the upwellings, comparable in magnitude to the mid-ocean ridge system and spaced about 1300 km apart. No such topography is discernible in Figure 2.4, and none has been convincingly demonstrated [129] by more sophisticated attempts to find it [130].

The proposal for convection driven by 'dripping' lower lithosphere also encounters the topographic constraint, but in this case the model implies only that there should be depressions where lower lithosphere is detaching [131]. The amplitude of such depressions has not been accurately estimated. Since a substantial amount of heat transport is required in order to 'flatten' the sea floor (about $40\,\text{mW/m}^2$ through old sea floor [99]), a significant and detectable amount of topography would be expected. Indeed, the higher viscosity of the cooler drips would enhance such topography by coupling their negative buoyancy more strongly to the surface. On the other hand, the elastic strength of the lithosphere would reduce the short-wavelength components of this signal. On balance, it is likely that there should be a network of depressions across the older sea floor, probably with amplitudes at least of the order of a few hundred metres. Such a signal should be readily observable, but it is not evident in Figures 2.4 or 8.4. Corresponding signals in the gravity field should also be present.

Some such signals have been demonstrated in restricted regions, but they appear to be due to something other than small-scale convection. The best-developed signals are in the southeast Pacific, where there are undulations of the sea floor with a wavelength of about 200 km and amplitudes of less than 200 m. Associated gravity and geoid anomalies have also been detected. The gravity anomalies are of low amplitude (5–20 mgal) and linear, with wavelengths of 100–200 km and lengths of the order of 1000 km [132]. Narrow volcanic ridges have also been found, coinciding with the gravity lows, and Sandwell and others [133] have argued that these are not compatible with small-scale convection. They propose instead that the lithosphere has been stretched over a broad region and that it has developed boudins, which are thinner, necked bands oriented perpendicular to the direction of stretching.

A further problem with the dripping lower lithosphere hypothesis is that its long-term effect would actually be to increase the rate of subsidence, not to decrease it as claimed. This is because it would enhance the rate of heat loss from the mantle, and thus would enhance the thermal contraction that is the primary reason for seafloor subsidence [134]. Previous conclusions had only taken account of the replacement of cool lower lithosphere by warm mantle, and had overlooked the influence of the cool lithospheric material as it sinks into the mantle.

The evidence for small-scale convection is thus absent or equivocal. As well, the claim that the old sea floor asymptotically approaches a constant depth, which motivated the idea of small-scale convection, is a misreading of the observations. We can conclude that, if any such mode of convection exists, other than the plate and plume modes, it must be a minor phenomenon that transports only a small amount of heat and generates only small geophysical signatures that are hard to resolve.

8.4 Other modes and causes 121

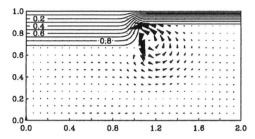

Figure 8.9. An example of edge convection in a numerical model. Lithospheric isotherms define a step in the lithosphere thickness. The cooler lithosphere projecting into the warmer mantle drives a downwelling adjacent to the step, which sets up a local circulation. The flow velocity in this model is 2.5 mm/yr. From King and Anderson [135]. Copyright Elsevier Science. Reprinted with permission.

8.4.4 Edge convection

A different possibility of small-scale convection is of so-called 'edge convection' [135]. This could arise where there is a step in the lithosphere thickness. Such a step would bring the cooler lithosphere of the thick part into proximity with the warmer mantle under the thin part. The resulting horizontal temperature gradient would drive convective flow. An example is shown in Figure 8.9. It has been suggested that such convection might occur, for example, at fracture zones that juxtapose lithosphere of different thicknesses, or at the edges of continents or the edges of Archaean continental nuclei.

King and Anderson [136] suggest that it might also occur during rifting, and this would increase the amount of melting as local mantle circulated through shallow depths. This is possible in principle, although their model is simplified and requires a number of significant assumptions. Also, they were motivated partly by the apparent inability of the plume head model to explain the volumes of melt in a flood basalt, but this problem has subsequently been resolved [76, 77], as explained in Section 7.4.

King and Ritsema [137] argue that edge convection around the South Atlantic region can explain hotspots and tongues of high seismic wave velocity, interpreted to be cold, located under the African and South American margins. This explanation of cold anomalies is plausible, but the connection with hotspots is not well established. The models are quite simplified and involve significant assumptions. Thus, although this idea may have some merit, a strong observational case is yet to be made.

8.4.5 Mantle wetspots

Green and Falloon [138] have argued that volcanic hotspots can be explained by mantle 'wetspots'. From a petrological point of view, this idea has some merits, since a small amount of water (less than 0.1%) can substantially reduce the solidus temperature, at which melting first occurs. It is also true that hydrated forms of minerals are generally less dense than their dry counterparts, which could provide the buoyancy required to explain hotspot swells. The effect on density needs to be better quantified, and it would need to be shown that observed water contents of hotspot volcanics are consistent with the amounts required to explain the buoyancy. It needs also to be shown that sufficient melt can be produced to explain the observed volcanism, since, although water reduces the solidus temperature, substantial degrees of melting still do not occur until the dry solidus temperature is approached.

However, a remaining difficulty would still be to explain the duration of long-lived volcanic centres like Hawaii. While a hydrated portion of the mantle, perhaps old subducted oceanic crust, might produce a burst of volcanism, there is no explanation offered for how the source might persist for 100 Ma or more. It is useful to estimate the volume of mantle required to supply the Hawaiian plume for 100 Ma. The total volume erupted into the Hawaiian and Emperor seamounts over 90 Ma is about 10^6 km^3. If we assume that there was about 5% melting of the source, this requires a source volume of 2×10^7 km^3, equivalent to a sphere of diameter 340 km. If such a large and buoyant region existed as a unit in the mantle, it would rise and produce a burst of volcanism. To explain the Hawaiian volcanic chain, the hydrated mantle material needs to be supplied at a small and steady rate.

The advantage of the thermal plume hypothesis is that a renewal mechanism is straightforwardly provided if the plume originates from a thermal boundary layer. It may be that the effects of water on melting and on plume buoyancy are significant, but it is far from clear that water alone could provide a sufficient explanation of the observations, while heat alone, or heat plus water, provides a straightforward and quantitatively successful account of the dynamical requirements of a theory of plumes.

8.5 Pursuing implications

This completes the discussion of mantle convection *per se*. With the general picture established, the rest of the book will explore implications. Potentially there are many implications, and it is a prime motivation for this book to provide enough understanding that these implications can be pursued more readily. However, I will focus on two that are of fairly fundamental importance, which I have been long involved with, and on which I am therefore reasonably well informed. These are the evolution of mantle convection and the chemistry of the mantle.

These topics are closer to the research frontier than the basic mantle convection covered so far. They will therefore be covered with more reference to the literature and perhaps assuming more basic knowledge of the subjects. Given the target audience of this book, this should not be inappropriate. Being closer to current research, they may also be regarded as less well established and more debatable. The main points of debate should be evident, and I make no apology for discarding interpretations that in my judgement are inadequate and developing new ones.

9

Evolution and tectonics

> Decaying heat sources and evolution of the system. Is there enough heat generation to keep the system cooking? Complications, and alternative controls on evolution. Implications for tectonic history. Archaean plates? Episodes?

Mantle convection is likely to have changed over the course of Earth's history. The mantle was probably hotter when the Earth formed, because of the large releases of gravitational energy during its accretion and the separation of the mantle and core. Also the radioactivity that drives the system now would have been stronger in past times, because of radioactive decay. Thus we might expect that mantle convection used to be more vigorous. But how much more vigorous? We can answer that question by building on the understanding we have already established.

It turns out that there may have been changes other than just a slowing down as the mantle cooled. For example, the melting and differentiation that accompanies mantle convection generate compositional density differences, and these may affect the course of the convection. Rheology may also have important effects. In fact there may have been some quite drastic changes in mantle convection, especially in the first half of Earth's history.

The geological record preserved in continents tells us that there have been changes in geological and tectonic processes over the past several billion years. Quite what the tectonic changes were is not yet very clear, but it is clear that the character of fragments of continental crust preserved from 3.5 Gyr ago is different in important ways from more modern crust. It seems also that the Earth's tectonic processes have proceeded in bursts, interspersed with quieter periods.

The ancient geological record is of course quite fragmented and incomplete, yet exploration continues, and chemical analysis techniques keep improving, extracting a surprising amount of information from tiny samples. Calculations of mantle convection will not be able to fill in the gaps and tell us definitively what was happening back then. Yet we should take advantage of all the information we have, and indeed a fruitful dialogue has developed among the various disciplines that study the ancient Earth.

Twenty years ago it seemed that studies of mantle convection might contribute a few basic points to the debates about ancient tectonics. For example, one might evaluate whether enough energy would have been available to drive a hypothesised tectonic process. However, the subject has matured surprisingly rapidly, and some particular hypotheses have been contributed to the dialogue, such as a mechanism that could cause major magmatic outbursts. It is therefore important to introduce this aspect of mantle convection.

9.1 Parametrised thermal evolution

If the temperature of the mantle changes, then its heat content will have changed. This will require heat to have entered or left the mantle. Reversing this logic, if we knew the rates at which heat was entering and leaving the mantle, we might calculate how fast it is warming up or cooling down. In words, we can write the equation

$$\text{rate of change of heat content} = \text{rate of heat input} - \text{rate of heat loss.} \quad (9.1)$$

If the rate of heat loss should exceed the rate of heat input, then the mantle would have a negative rate of change of heat content. In other words, it would be cooling.

Continuing this line of thought, if we know how fast the mantle is cooling now, we can estimate how hot it was some time in the past, say one billion years ago. However, one billion years ago some things might have worked differently. For one thing, the rate of radioactive decay of key elements would have been higher, because they would not have decayed as much as they have now. For another, if the mantle were hotter, it might convect faster. We can say how much faster the radioactive decay was back then. Can we also say how much faster mantle convection might have been? Yes, we can. It's implicit in our formula for the convective velocity that we developed back in Chapter 5. We can also calculate what the rate of heat loss would have been. If we had the rate of heat input and the rate of heat loss that applied one billion years ago, then we could calculate a new rate of cooling that applied back then. Perhaps it was larger than now. We could then repeat the exercise, and estimate what the mantle temperature was two billion years ago. And so on.

That, essentially, is how one can calculate what the temperature of the mantle was in the past. From the mantle temperature we can get its convective velocity, which might be the velocity of tectonic plates back then. We might also get its rate of heat loss, and thus some basic notion of what level of tectonic activity might have been occurring. In this section we will show how that calculation works in a little more detail. Later we will look at potential complications, such as density differences caused by melting.

To begin, let's relate the rate of change of heat content to the rate of change of temperature. Recall, from Section 5.4, that the change in heat content, ΔH, is related to the change in temperature, ΔT, through the specific heat, C_P:

$$\Delta H = M C_P \Delta T, \tag{9.2}$$

where M is the mass of the mantle. If these changes occur within a time interval Δt, then we can relate the *rates* of change:

$$\frac{\Delta T}{\Delta t} = \frac{1}{M C_P} \frac{\Delta H}{\Delta t}. \tag{9.3}$$

There are two kinds of heat input into the mantle: radioactive heating, Q_R, and heat entering from the core, Q_C. Heat is being lost through the Earth's surface, Q_S. Thus

$$\frac{\Delta T}{\Delta t} = \frac{1}{M C_P}(Q_R + Q_C - Q_S). \tag{9.4}$$

This expresses the same relationships as Eq. (9.1).

To go further, we need detailed expressions for the heat inputs and loss. Most of the radioactive heating of the mantle is due to four isotopes: ^{238}U, ^{235}U, ^{232}Th and ^{40}K. The decay of these isotopes is well determined, but the expressions involved are a little messy, so they are given in Appendix B. The abundance ratios of the isotopes are fairly well determined, but the absolute amounts are still debated, as we will see later. For this reason, we define an adjustable ratio, the Urey ratio, as the ratio of present heat generation in the Earth to present heat loss, i.e.

$$\mathrm{Ur} = Q_R/Q_S = M H_R/Q_S, \tag{9.5}$$

where H_R is the heat generation per unit mass given by Eq. (B.1) in Appendix B. In the tradition of fluid-dynamical ratios like the Rayleigh number, the Urey ratio is given a two-letter symbol, Ur, so the 'r' is not a subscript.

In Chapter 5 we derived an expression for the velocity of plates, Eq. (5.18), and an expression for the heat flow out of the mantle, Eq. (5.22). If we leave out the approximate numerical factor and use previous expressions for Ra and q_c, Eq. (5.22) can be written as

$$q_S = \frac{KT}{D}\left(\frac{g\rho\alpha T D^3}{\kappa\mu}\right)^{1/3}, \tag{9.6}$$

where the subscript on q_S is to remind us that this is the surface heat loss per unit area of the Earth's surface. Then

$$Q_S = A_E q_S = 4\pi R_E^2 q_S, \tag{9.7}$$

where A_E is the area of the Earth's surface and R_E is the radius of the Earth. Temperature occurs explicitly twice in Eq. (9.6), and implicitly once, because the viscosity is a strong function of temperature, as we saw in Chapter 4 (Eq. (4.9) and Figure 4.4). The other quantities involved in Q_S are either constant or not likely to vary much as the mantle temperature evolves. Thus, although these equations look rather messy, the essential point is that they define how the heat loss varies as the mantle temperature varies. Thus, referring back to the initial discussion, we could estimate what the heat loss was one billion years ago, when the mantle was probably a bit hotter than at present.

An analogous expression can be derived for the heat transported by mantle plumes, though it's even messier. It can be found in *Dynamic Earth* [1]. The essential points here are that it is a little less sensitive to temperature, and that the relevant viscosity is the viscosity of the hottest mantle material, right next to the core. That's because it has the lowest viscosity and is the most mobile material.

Thus we can get expressions for Q_R, Q_C and Q_S in Eq. (9.4) that depend on the temperature of the mantle or on time. That means we can do the calculations outlined at the start of this section, i.e. calculate the present rate of change of temperature, estimate from that the mantle temperature at some previous time, recalculate the heat terms and the rate of change of temperature, estimate the temperature at an even earlier time, and so on. This, essentially, is what is done, except it's done numerically and using much smaller time steps than one billion years. There are well-documented clever ways of minimising errors in such a calculation, but we don't need to go into all that. You can find everything you ever wanted to know in *Numerical Recipes* [139], an excellent resource.

The temperature of the core can be calculated in a similar way. The heat loss from the core is just Q_C, and this calculation assumes there is no radioactivity in the core, though that is debated by some. Thus the relationship is like Eq. (9.4) with only the heat loss term, and with parameters appropriate for the core. Both of these calculations can also be run forwards in time, if you assume a starting temperature.

The result of one such calculation is shown in Figure 9.1. The calculation was run forwards in time. The initial state is envisaged to be like that sketched in Figure 7.3, curve (a), in which the mantle and the core have the same temperature where they contact. Parameters have been adjusted to match observed present values of the upper-mantle temperature (T_U), surface heat flow (Q_S) and plume heat flow (Q_C). The lower-mantle temperature (T_L) is assumed to be 1000 °C higher than the upper-mantle temperature at present. The core temperature (T_C) is assumed to

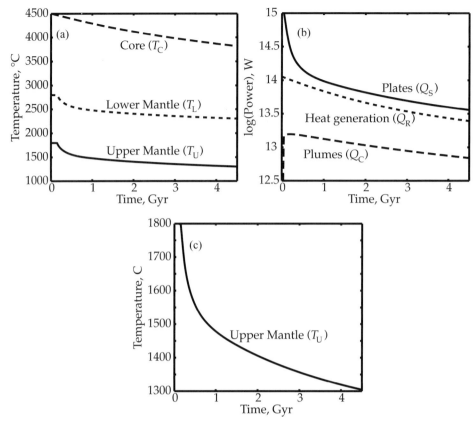

Figure 9.1. A reference thermal evolution of the mantle and core. Details of the parameters and outputs are given in Section B.2.

have started equal to T_L. It was necessary to adjust the Urey ratio to match the present surface heat flow, the efficiency of plate heat flow to match the present mantle temperature, and the efficiency of plume heat transport to match the present plume heat flow. The 'efficiency' factors are effectively the not well-determined numerical factors in the formulae for Q_S and Q_C.

The thermal evolution in Figure 9.1(a) looks fairly simple: everything declines monotonically. However, there are a number of features to comment on. There are two phases of cooling, an early rapid cooling lasting about 500 Myr and then a much slower cooling. The transient early phase occurs because the assumed initial mantle temperature is higher than the temperature that would be maintained by the radioactive heating. This high starting temperature was chosen to reflect the high temperature expected to have been generated by the formation of the Earth and the separation of the mantle and core materials. We don't have an accurate estimate of what that early temperature might have been, but the calculation makes it clear that any initial excess temperature would decay away within a few hundred million years, and it's useful to know that.

The early transient cooling also establishes the temperature difference between the core and the lower mantle. This temperature difference is required to drive mantle plumes, and Figure 9.1(b) shows the plume heat flow, Q_C, rising from zero at the start. This quantifies the evolution sketched in Figure 7.3, where it was argued that plumes are inevitable if the core is hotter than the mantle. After the early transient cooling, the mantle temperature is maintained by radioactive heating. You can see this from Figure 9.1(b), where the heat loss, Q_S, drops rapidly until it is just above the heat generation, after which it tracks the heat generation fairly closely. Q_S is always a little higher than Q_R because the mantle is cooling. (If Q_R were constant instead of declining, and there were no core heat flow, then Q_S would approach it asymptotically.)

Figure 9.1(c) shows the mantle temperature on an expanded scale, and it declines by only about 200 °C in the slow cooling phase, although the heat loss declines by a factor of 3–4. The reason the temperature does not decline very much is because the mantle viscosity depends so strongly on temperature (Chapter 4). It takes only a modest change in temperature to increase the viscosity substantially, and thus to slow the convection and reduce the heat loss substantially.

The core temperature declines only slowly, reflecting the low heat loss, Q_C, which is controlled not by the core but by the thermal boundary layer at the base of the mantle, which regulates how much heat can escape through the mantle. Q_C is controlled mainly by two quantities, the temperature difference between the mantle and the core, and the viscosity within the thermal boundary layer, which is controlled by T_C. Because T_C does not decline very much, the heat carried by plumes does not vary a lot through the age of the Earth in this calculation (Figure 9.1(b)). It also happens that $(T_C - T_L)$ declines only slowly in this calculation, which contributes to Q_C declining only slowly: it peaks at 16 TW and declines to 7.6 TW at present. Thus this calculation suggests that plumes may have been only moderately more active in the past than they are now.

We will use this simplified thermal evolution model as a reference case, as we discuss potential variations and complications below. Because of its simplifications, it requires some parameters to specify simple results of complicated processes. For example, when Eq. (9.6) is used, an additional adjustment factor is included (Section B.2) to compensate for the simplified approximations used in its derivation. This approach has therefore become known as parametrised convection, as distinct from the numerical convection (below) in which the physics is left freer to work itself out.

For the moment, the parametrised model suggests some significant lessons. If the Earth started hot, due to the gravitational energy released during accretion and core formation, then that 'primordial heat' would have been removed within a few hundred million years. Or, rather, the heat would have been reduced to the level maintained by radioactive heating (heat is heat, and there is of course no distinction between the initial heat and the radiogenic heat added later). Since then, the mantle's temperature and convective activity would have been controlled by radioactive heating. Thus there is no thermal 'vestige of a beginning' in the mantle, although of course vestiges may have been left in rocks surviving from that time. The rapid early cooling is also a result of the strong dependence of viscosity on temperature, because a few hundred degrees higher temperature lowers the viscosity by orders of magnitude and thus speeds convection and cooling.

The slow cooling since the early transient reflects the slow decay of radioactive heating. The relatively small temperature change is a significant result, because it says that we should not expect the mantle to have been radically hotter in the Archaean. However, plate velocities might have been 10 times greater than at present during the Archaean. This is because velocity varies as the square of heat flow, as is implied by Eqs (5.23), and heat flow has dropped by a factor of about 3. We will see this explicitly below, from a numerical model. Thus tectonic activity might have been substantially higher.

This calculation quantifies the argument given in Section 4.4, due to Tozer, that mantle convection is fairly inevitable in an Earth-like planet. If Earth started hot, it will convect rapidly and cool until its rate of convection is sufficient to remove the heat generated by radioactivity. If the Earth started cool, the mantle would not convect at a significant rate, so radioactive heat would accumulate and, given enough time, it would reach a temperature at which convection would become vigorous enough to remove the radiogenic heat.

A basic question about Earth's history is whether plate tectonics has operated throughout Earth's history. There seem to be no decisive arguments either way at present, substantially because of the limited geological record from early times, as we will see later. A more general question is whether the mantle convected in such a way that heat removal was governed by the kind of relationship shown in Eq. (9.6). This depends on how the upper thermal boundary layer behaved, since it controls mantle cooling and at present it comprises a set of stiff but mobile plates. Even if there were no plates, Eq. (9.6) might still apply if the boundary layer was mobile. On the other hand, its mobility might be limited by rheological or compositional factors and then our simple thermal evolution might not apply. This point will be taken up later in this chapter.

9.2 Numerical thermal evolution

Within the last decade computer power has advanced enough to enable the high resolution required for numerical models of mantle convection when the mantle

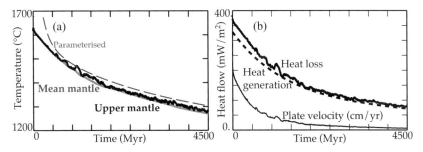

Figure 9.2. Summary of the thermal evolution of a numerical model of mantle convection with decaying heat sources. (a) Upper-mantle temperature (solid black), mean mantle temperature (solid grey) and the temperature from the parametrised calculation in Figure 9.1 (dashed grey). (b) Surface heat loss (thick solid) and radioactive heat generation (thick dashed). The plate velocity (thin solid) is included.

was hotter and thermal boundary layers thinner, and to run the models for the age of the Earth, at least for two-dimensional calculations. Results from a model in which radioactive heat generation decays appropriately are shown in Figures 9.2 and 9.3. Details and parameters of this model are given in Section B.3.

Figure 9.2(a) includes the temperature curve from the parametrised calculation in Figure 9.1. They are quite similar once the effect of the different starting temperatures passes, and apart from a small vertical offset that reflects the difficulty of predicting exactly the temperature at which the numerical model will run. This reflects the similar radioactive decay and similar temperature dependence of viscosity used in the two models. This similarity gives some confidence in both kinds of calculation.

As before, the surface heat loss (Figure 9.2(b)) tracks the heat generation closely – more closely in this case because there is no core heat included in this model, so radioactivity provides the only heat input.

Figure 9.2(b) includes the velocity of the surface plates in this model. Their present velocities are 6.2 cm/yr but they were much higher in the past. In the Archaean (0.7–2 Gyr in the plots) they were 18–60 cm/yr. The thermal boundary layer and the subducting plates are correspondingly thinner at the earlier times (Figure 9.3). Subduction is not modelled with great accuracy in this model, but the impression of slower and thicker plates and subducted lithosphere is well conveyed in Figure 9.3. The slowing plates and the corresponding slowing of convection velocities within the mantle are also evident in this figure because the spacing of the streamlines is inversely proportional to the local velocity.

Returning to Figure 9.2, mean temperatures for both the upper mantle and the whole mantle are included in panel (a). In this model the lower mantle is 30 times more viscous than the upper mantle, as indicated by observations (Chapter 4). There

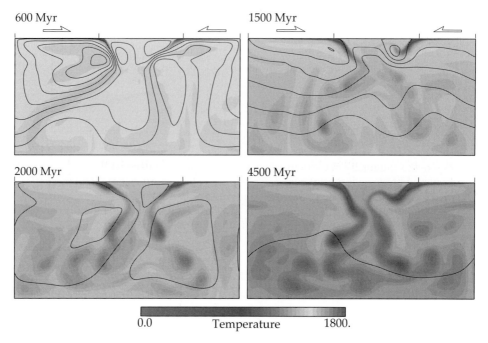

Figure 9.3. Temperature and flow field at selected times from the evolution shown in Figure 9.2. Temperature is shown in the grey scale (only in the 600 Myr panel of this model is there any fluid, under the plates, hotter than the palest grey). The flow field is represented by streamlines, which are the flow lines for that moment. The surface has three plates, two outer oceanic plates subducting under the central continental plate. Their motions are indicated by the arrows. The continental lithosphere is excluded from this model, so the middle section of the top surface corresponds to the base of the continental lithosphere.

are short-term fluctuations in the upper-mantle temperature that reflect the higher velocities there, and also the vagaries of how quickly the subducted lithosphere sinks into the lower mantle. These fluctuations are also closely reflected in the heat loss (panel (b)). The mean mantle temperature, on the other hand, varies much more smoothly, as it responds to the time-integrated heat loss.

9.3 The heat source puzzle

The calculations in the previous sections implicitly assume that there is sufficient radioactive heating to sustain the present mantle heat loss, allowing for the fact that the mantle is cooling. At the end of the parametrised run, the heat loss is 36.4 TW, whereas the radiogenic heating is 24.6 TW (Section B.2). Thus the Urey ratio is 0.68. In other words, 68% of the surface heat loss is supplied by radiogenic heating.

9.3 The heat source puzzle

The remaining 30% comes from two sources: the core (7.6 TW) and internal heat released as the mantle cools (the balance, 4.2 TW).

We can check this by looking at the rate of temperature decrease implied by the 4.2 TW released by cooling, using Eq. (9.3). Taking the mantle mass to be 4×10^{24} kg and its specific heat to be 1000 J/kg °C, the temperature declines at a rate of 1.05×10^{-15} °C/s = 33 °C/Gyr. This corresponds well with the decline of 31 °C over the final billion years of the model.

The silicate part of the Earth is estimated, from meteorite studies, to contain about 20 ng/g of uranium [140]. (I give concentrations in this form, nanograms of U per gram of rock, to avoid the ambiguity of the term 'parts per billion' commonly used by geochemists, because the latter could be a ratio by weight, by volume or by moles.) The ratios of thorium and potassium to uranium are commonly estimated to be about Th/U = 3.8 g/g and K/U = 1.3×10^4 g/g (Section 10.2). These quantities will generate about 5 pW/kg of heat [141] (1 pW = 1 picowatt = 10^{-12} watt). With a silicate mass of about 4×10^{24} kg this implies a total heat production of 20 TW. This is a little less than the radiogenic heating required by the thermal evolution model, but the difference is probably within the uncertainties of the model and the geochemical estimates.

However, there is a problem, because about half of the Earth's heat source elements are estimated to have been sequestered into the continental crust [142]. The heat generated within the continental crust (and lithosphere) will conduct directly to the surface and play no part in heating the mantle. Because the continental crust is quite heterogeneous, estimates of its uranium content have varied through the range 30–60% of Earth's budget [118]. Thus only 40–70% of the Earth's total uranium is available to heat the mantle, so it would generate only 8–14 TW of heat. With a total mantle heat loss of about 35 TW, this implies a Urey ratio of only 0.23–0.4, much less than the 0.7–0.9 range required by the thermal evolution models.

This is a major discrepancy that implies there is something important we don't understand about the Earth. There are three possible resolutions of the puzzle. First, the geochemical estimates could be wrong. Second, the present rate of heat loss is unusually high, with the implication that the present rate of cooling is also unusually high. Third, the rate of heat loss and the rate of cooling are not unusually high, with the implication that the mantle was a lot hotter in the past than implied by the thermal evolution calculations just presented. The third option would also imply that there is something important missing from the physics of mantle convection as so far presented.

The first option will be discussed in Chapter 10. For now we can note that there are certainly uncertainties in both the values and the assumptions involved in geochemical estimates, but a higher uranium content of the Earth implies other puzzles about Earth's composition.

In the second option, the transient cooling might have lasted for several hundred million years without implying excessively high mantle temperatures in the past, but any longer would raise questions about geological processes. If the mantle has only 10 TW of radiogenic heating plus 7 TW from the core, and a mantle heat loss of 35 TW, then 18 TW has to come from cooling. This implies a cooling rate of about 140 °C/Gyr, compared with the cooling rate in the models of the previous sections of around 35 °C/Gyr. If such a cooling rate had been sustained for 500 Myr, then the mantle would have been 70 °C hotter in the Palaeozoic than now. A temperature that high might imply magmatism that ought to be evident in the geological record, but our calibrations of the mantle models and of mantle petrology are not so secure as to be able to strongly rule out the possibility.

Some fluctuations in the cooling rate are expected to have occurred just due to normal fluctuations in the age–area distribution of oceanic plates [143]. However, these may not have a large enough amplitude [144], or might imply higher rather than lower heat flow in the recent past [145]. A more interesting possibility is that plate tectonics might have been intermittent, as suggested by Silver and Behn [146, 147]. This would cause larger fluctuations in heat loss, and would allow the mantle to maintain its temperature over the longer term despite the present imbalance.

The third option is that the heat imbalance is long term and the mantle has been cooling more rapidly for a long time. A mechanism that might have this result has been advocated by Korenaga [148, 149]. Korenaga argues that plate velocities are limited not just by the viscous resistance to flow that we invoked in Chapter 5, but by the energy it takes to bend the lithosphere as it enters a subduction zone. He argues as well that the plate thickness is controlled not just by conductive cooling but also by small-scale convection under the plate and by the dehydration that accompanies melting under mid-ocean ridges, which increases the local viscosity. The latter assumption leads to plates that are thicker in a hotter mantle, rather than thinner. This increases the bending resistance to the plates, which slows them and prevents the heat loss from being large. The result can be a mantle temperature that peaked around 3 Gyr ago. An example of this kind of evolution is shown in Figure 9.4 (black curves). This solution assumes a radius of curvature of subducting plates of 200 km.

Solution '200' has a present cooling rate of 200 °C/Gyr and a present Urey ratio of 0.3. Early in Earth history the heat loss (Figure 9.4(b)) is less than the heat generation (plus heat released by core cooling, not shown) and the mantle warms. The (upper) mantle temperature peaks at about 1750 °C about 3.4 Gyr ago, at which time the heat loss and plate velocities reach minima because the plates reach their

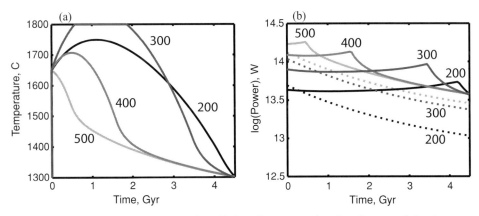

Figure 9.4. Thermal evolutions in which resistance to plate bending at subduction zones limits plate velocity. Curves are labelled with the assumed bending radius of the plate. (a) Temperature. (b) Heat loss (solid) and heat generation (dotted). From Davies [146]. Copyright Elsevier Science. Reprinted with permission.

greatest thickness, due to the greater melting and dehydration. Thereafter the heat loss slowly increases and heat inputs decline. At about 300 Myr in this particular solution the dehydration thickness becomes less than the thermally defined thickness of a plate, and the solution briefly reverts to conventional behaviour, with the heat loss dropping rapidly as the mantle cools further.

This is an innovative and interesting evolution, but it depends on assumptions about several phenomena that are not very well constrained, including melting depths and dehydration at high mantle temperatures, and the radius of curvature of subducting plates. It turns out to be very sensitive to the bending radius of plates. Observations of present plates indicate bending radii ranging from 120 km to 1200 km, with most between about 160 km and 600 km [150]. The effect of assuming larger radii is shown in Figure 9.4. With radii of 400–500 km, the bending resistance still dominates early in Earth history, but then it falls below the viscous resistance of thermal plates and the solution reverts to the behaviour of conventional models. The present Urey ratio is 0.8 and the cooling rate is about 30 °C/Gyr for both of these solutions. The '300' solution is transitional, but even in this case the present Urey ratio is 0.67, well above the value required to reconcile long-term heat loss and heat generation. Temperatures in this solution exceed 1800 °C in the Archaean, which surely is excessively high.

Thus this proposed reconciliation of heat generation and heat loss seems to require bending radii that are close to the extreme observed today, and that would need to have applied even in the past when the plate thickness, defined by melting and dehydration, would have been even greater. The implausibility of this is increased by the modelling of Capitanio *et al.* [151], which shows that plates tend to self-regulate their bending radius so the bending resistance does not become dominant. There would in any case be questions about the consistency of this kind of thermal history with the geological record, as it would seem to imply rather extreme levels of magmatism and tectonics in the Archaean. This, however, is less definitive than the implausibility of the parameters required.

At present it thus seems that there is no plausible proposal for how a long-term imbalance of heat loss and heat generation might be maintained. We are left with the other two possibilities, that the geochemical estimates are wrong or that we are in a period in which the heat loss is unusually high. Either of these options seems possible. A little more will be said as we go along, but the resolution of the heat source puzzle will remain for the future.

9.4 Compositional buoyancy

The convection theory of Chapters 5 and 6 and the thermal evolution calculations of this chapter so far assume that thermal buoyancy is the dominant driver of mantle convection. Yet we know that plate tectonics generates density differences due to different compositions. Melting at mid-ocean ridges forms basaltic oceanic crust, and it also leaves a zone of mantle depleted of its basaltic component to a depth of around 60 km. The depleted mantle tends to be buoyant because the erupted basalt is more enriched in iron than its source, according to conventional views. This means that subducted lithosphere is not uniform – it is stratified in composition and in density.

The oceanic crust is also buoyant relative to the mantle, having a density of about 2900 kg/m^3, compared with a mantle density of 3300 kg/m^3. However, by a depth of about 60 km it is expected to have transformed to a different mineral assemblage, called eclogite, with a density around 3500 kg/m^3. Mantle minerals undergo further phase transformations as they go further down, because of the increasing pressure. There is a significant transformation around 410 km depth and a larger one around 660 km depth. Subducted oceanic crust undergoes comparable transformations with depth, and generally it is a little denser than average mantle at the same depth, by 100–200 kg/m^3 [152, 153]. There is, however, a depth interval within which it is less dense than adjacent mantle, because its transformation to lower-mantle phases is delayed until around 750 km depth. Thus between 660 km and 750 km depth it is still in its transition zone phases and is less dense by about 150 kg/m^3.

The density difference between oceanic crust and the mantle, and its several changes of sign, has some important effects on mantle dynamics. This topic is still being explored, but there are some important conclusions or indications that will be outlined here.

9.4 Compositional buoyancy 137

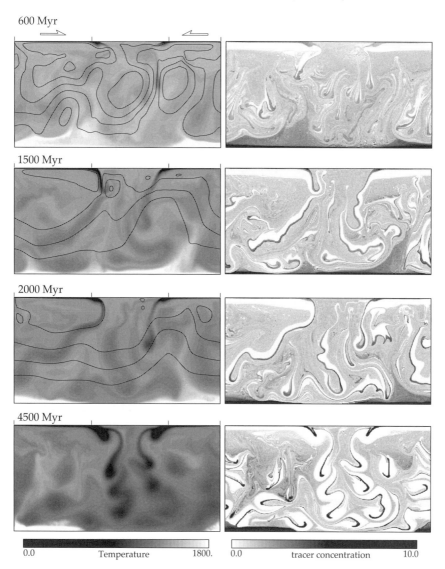

Figure 9.5. Temperature and streamlines (left) and tracer concentrations (right) from a numerical model with tracers that represent subducted basaltic crust that is denser than average mantle by 150 kg/m^3. A tracer concentration of 1 corresponds to the average basaltic content of the mantle. White zones at the top are melting zones from which tracers have been removed to the oceanic crust.

9.4.1 Four billion years of basalt subduction

We will start by looking at the effect of subducted oceanic crust being (mostly) denser than surrounding mantle. Results from a numerical model are shown in Figure 9.5. The model represents oceanic crust (at the surface or within the mantle)

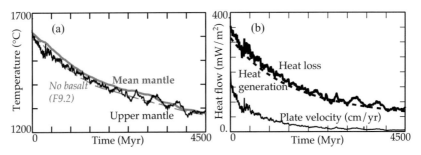

Figure 9.6. Thermal evolution of the model depicted in Figure 9.5. The mean mantle temperature from a model with no basaltic tracers (Figure 9.2) is included for comparison.

using tracers that each carry a small mass corresponding to the negative buoyancy of a piece of oceanic crust. Details of the model, including how the basalt tracers are implemented, are given in Section B.4.

As tracers rise into a defined melting zone, they are removed and placed in a thin oceanic crust. This simulates melting under mid-ocean ridges and results in the formation of compositionally heterogeneous lithosphere. This subducted lithosphere can be seen in the right panels as the white (empty) zones bordered by a thin sheet of concentrated tracers.

The figure shows that some of the tracers settle to the bottom of the model and accumulate there. Tracers are also removed from accumulations and mixed back into the mantle interior by warm upwellings. Over time, a relatively steady number of tracers are located in the bottom accumulations, though the number declines slowly as the model evolves. The zone where the tracers accumulate has a higher density that tends to resist being mixed up into the mantle interior. Because of its limited circulation, heat accumulates within it, and it becomes several hundred degrees hotter than the mantle above (left panels). This zone of dense but hot mantle raises the average temperature of the mantle, as can be seen in Figure 9.6, which shows the thermal evolution of this model. The mean temperature of this model is significantly higher than for the model in Figure 9.2 that lacked basaltic tracers.

The tendency of subducted oceanic crust to accumulate at the base of the mantle was first demonstrated by Christensen and Hofmann in 1994 [122]. The improvement in computer power since then permits models that have greater realism, including running at the full mantle Rayleigh number and evolving for the full age of the Earth.

The basaltic accumulations have significant structure within them (Figure 9.5, right panels). There tends to be a thin zone, 50–100 km thick, with concentrations two or more times the average. Above that zone are regions of lesser and variable concentration that extend upwards for hundreds of kilometres. The shapes of these zones are quite variable, as they are pushed around by sinking subducted lithosphere. This picture corresponds quite well in general terms with the picture from seismology of a thin D″ zone, perhaps 200 km thick, and larger anomalies extending hundreds of kilometres upwards under Africa and the western Pacific, as described earlier (Figure 8.8).

This kind of model also yields important geochemical results that will be taken up in Chapter 10. Indeed, that was a primary motivation of Christensen and Hofmann's original models. This model, and others shown here, assume the operation of plate tectonics throughout the Earth's history. It is not known if plate tectonics really did operate in the first half of Earth history, but any behaviour of the top thermal boundary layer that formed and recycled a mafic crust would probably yield mafic accumulations like those found here. One can treat the assumption of plate tectonics as a hypothesis. If it is not true, then discrepancies with evidence ought to emerge. In the meantime it yields well-defined models that provide a benchmark for alternative hypotheses.

9.4.2 Buoyancy of oceanic crust

The oceanic crust is significantly less dense than the mantle within about 60 km of the surface. As lithosphere begins to subduct, the compositional buoyancy of the oceanic crust will counteract some of the thermal negative buoyancy of the lithosphere. Except for lithosphere younger than about 20 Ma, the negative thermal buoyancy dominates and the lithosphere can subduct. The implications are fairly minor for the present Earth. Presumably any lithosphere younger than about 20 Ma that subducts must be pulled down by attached older lithosphere, or by other forces in the system.

However, in the past the implication might be very important. If the mantle is hotter, then there ought to be more melting under mid-ocean ridges and that should produce thicker oceanic crust. At the same time, convection will be faster in a hotter mantle and plates will be younger, on average, when they reach subduction zones. Thus the negative thermal buoyancy is lower and the positive compositional buoyancy is greater. This means that plates will have to be older than 20 Ma before their net buoyancy becomes negative. Positively buoyant plates cannot subduct. They would have to age further until they were able to subduct. Thus plate tectonics might still operate, but more slowly than required to cool the mantle. This could lead to a thermal runaway in which the mantle gets hotter in time, and this leads to the paradox that its present cool state cannot be explained. Thus, presumably, something else would have to have happened.

This argument was made in 1992 [154] and for some time seemed to preclude plate tectonics before about 2 Gyr ago. However, models like that in Figure 9.5

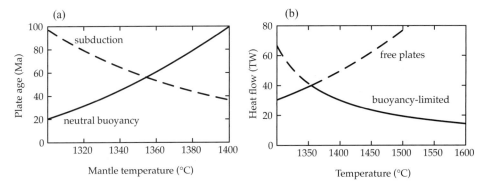

Figure 9.7. (a) Age at which plates become neutrally buoyant, and average age of plates as they reach a subduction zone. (b) Heat removed by free plates and by plates limited by compositional buoyancy versus mantle temperature. Actual heat loss would be given by the lower (solid) parts of the curves. After Davies [154]. Copyright by the Geological Society of America.

demonstrate another phenomenon that seems to overcome the argument. The argument and its antidote will be summarised.

Davies [154] estimated the positive buoyancy contributed by both the oceanic crust and the depleted melt zone from which the crust is drawn and calculated the age at which lithosphere becomes neutrally buoyant. Using estimates of how melting increases as mantle temperature increases, the age of neutral buoyancy was calculated as a function of mantle temperature, and is shown in Figure 9.7(a). The theory of convection from Chapter 5 can be used to calculate plate velocity as a function of mantle temperature. The plate velocity determines how old a plate is when it reaches a subduction zone, and Figure 9.7(a) also shows the average age at subduction versus mantle temperature.

The curves show that, when the mantle was only about 50 °C hotter than now, plates would be neutrally buoyant as they reached subduction zones. At higher temperatures, plates would not be able to subduct if they moved at speeds calculated from the convection theory. According to Figures 9.1 and 9.2 the neutral buoyancy condition would have applied only about 1.5–2 Gyr ago.

Before that time, plates would have had to move more slowly, so heat removal by plates would have been less than predicted by free-plate theory. This is illustrated in Figure 9.7(b). At a little higher temperature again, heat loss would be less than radioactive heating, so the mantle would have been warming up rather than cooling down. If that were true, then it would never have been able to cool to the neutral point anyway – a paradox. The only way out of the paradox is to assume that some other heat removal mechanism could operate. It is possible to envisage different behaviour of the cool thermal boundary layer, and hence a different tectonic regime. Discussion of such possibilities will be deferred to Section 9.6.

9.4 Compositional buoyancy   141

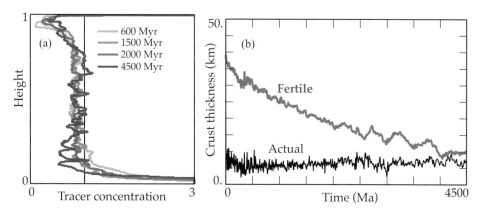

Figure 9.8. (a) Horizontally averaged tracer density versus height in the model of Figure 9.5. There is a gradient of tracer content in the upper mantle, representing a gradient in basaltic content. (b) Thickness of the oceanic crust calculated from the same model (Actual), and what it would have been if the uppermost mantle had average fertility, i.e. average basaltic content (Fertile).

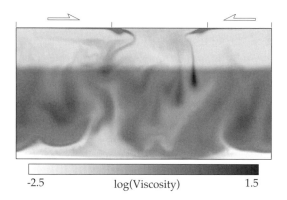

Figure 9.9. Viscosity at 600 Myr in the model of Figure 9.5. Viscosity depends on temperature, hotter fluid having lower viscosity. Superimposed on this is a depth dependence in which the viscosity of the lower mantle is 30 times the viscosity of the upper mantle.

More recent results have revealed a different way out of the paradox [123, 155]. Figure 9.8(a) shows a gradient of tracer concentration in the upper mantle of the model in Figure 9.5. This occurs because the upper mantle has a lower viscosity, especially earlier in Earth history when it was hotter (see Figure 9.9). This lower viscosity offers less resistance to sinking tracers, which settle through the upper mantle and then (mostly) remain in suspension in the lower mantle, which has a viscosity 30 times higher. This implies that the uppermost mantle would be depleted in its basaltic component, because old oceanic crust would tend to sink deeper into

142   *Evolution and tectonics*

the mantle. The uppermost mantle would therefore be less fertile, produce less melt, and result in a thinner oceanic crust.

(There are minima in the curves in Figure 9.8(a) that correspond with the melting zone, but the gradient below that is clear. The results in Figure 9.8(b) are a more direct and reliable measure of the depletion.)

Figure 9.8(b) shows the crustal thickness calculated from the model, based on the number of tracers removed from the melt zone into the oceanic crust ('Actual'). Remarkably, it averages only 5–8 km and is never greater than about 10 km, being smaller in the past. For comparison, the crustal thickness that would be generated by mantle of normal basaltic content is also shown ('Fertile'). The fertile thickness declines steadily from almost 40 km. The difference between these two curves is a measure of the degree of depletion relative to the present mantle fertility, and it shows a depletion by a factor of 3–6 early in Earth history. Other models of this type have shown comparable degrees of depletion [123, 155, 156]. This has important geochemical implications as well, which will be discussed in Chapter 10.

The implication of this result is that the buoyancy of the oceanic crust may not have been an important factor inhibiting subduction. Thus the possibility that plate tectonics has operated through most of Earth history is revived. Of course, there may be other reasons why plate tectonics did not operate, but this difficulty does not seem to be as serious an obstacle as it first appeared.

9.4.3 *A mid-mantle basalt barrier and early mantle overturns*

The other depth range in which oceanic crust is believed to be less dense than average mantle is below the transition zone, 660–750 km depth. In this depth range, it will therefore tend to rise, whereas in the upper mantle it tends to sink. Thus there will be a tendency for old subducted oceanic crust to accumulate around 660 km depth. This turns out to have strong effects early in Earth's history.

An example is shown in Figure 9.10. In the first two panels of this model, tracers of subducted oceanic crust have accumulated around 660 km depth to such an extent that they have blocked vertical flow, and the convection is separated into two layers. There is a large temperature difference between the layers. These features are not present in the later two panels.

A fuller picture of the behaviour of this model is conveyed in Figure 9.11, which shows summary plots of the evolution of the model. Immediately evident are very large swings in the upper-mantle temperature (panel (a)). The first billion years of these swings are shown on an expanded timescale in panel (c), and it is apparent that the upper-mantle temperature makes sudden jumps upwards, followed by smooth declines leading into another jump.

9.4 Compositional buoyancy

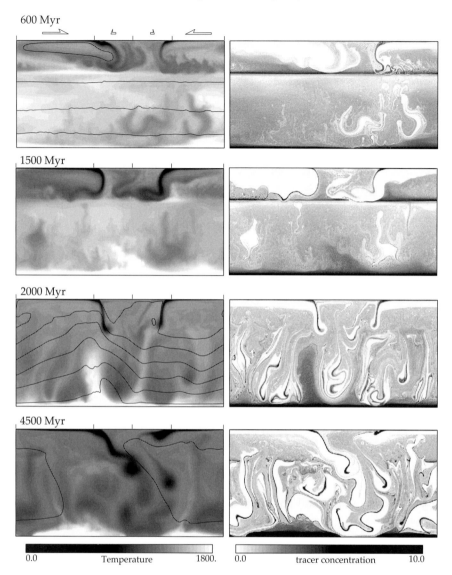

Figure 9.10. Evolution of a model in which subducted oceanic crust is buoyant in the depth range 660–750 km. The subducted crustal material accumulates around 660 km to form a 'basalt barrier' that separates mantle convection into two layers in the first two panels. After Davies, Case 2 [156]. Copyright Elsevier Science. Reprinted with permission.

The swings occur because the layering breaks down episodically, allowing hot fluid from the lower mantle to flood into the upper mantle, suddenly raising its temperature. This is confirmed in Figure 9.12, which shows a breakthrough in progress. Layering is soon re-established, after which the upper mantle cools smoothly but

144 Evolution and tectonics

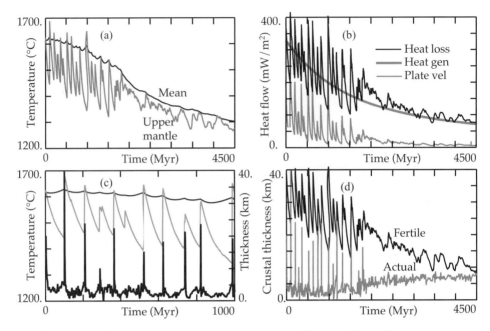

Figure 9.11. Summary of evolution of the model of Figure 9.10. (a) Upper-mantle and mean mantle temperature. (b) Heat flows and plate velocity. (c) Detail of temperatures (panel (a)) and crust thickness (panel (d)). (d) Thickness of the oceanic crust. After Davies [156]. Copyright Elsevier Science. Reprinted with permission.

fairly rapidly. The breakdowns evidently occur when the temperature difference between the two layers becomes so large that their density difference overcomes the basalt barrier.

The surface heat loss and plate velocity vary in concert with the upper-mantle temperature (Figure 9.11(b)), as does the putative 'Fertile' thickness of the oceanic crust (Figure 9.11(d)). However, the variation of the actual crustal thickness shows dramatic spikes during overturns, with thicknesses of only 2–4 km between overturns. The 'Actual' crustal thickness is included in Figure 9.11(c) on an expanded timescale, and it can be seen that the spikes coincide closely with the mantle overturns. This is the expected result of very hot and more fertile lower-mantle fluid flooding into the upper mantle. During overturns, the crust reaches thicknesses of 10–30 km.

Such crustal thicknesses represent a major magmatic episode, enough to cover the entire Earth in tens of kilometres of basalt. These bursts last only a few million years. The models of melting are not really realistic during overturns, because plate tectonics is assumed to continue through them, whereas such thick oceanic crust would surely stop subduction for a period. However, the models are taken to be

9.4 *Compositional buoyancy* 145

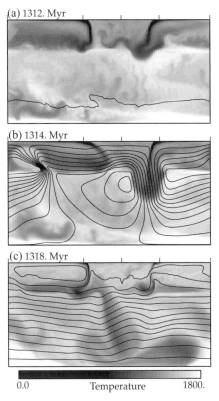

Figure 9.12. Detail of an overturn in the model of Figure 9.10. Cool fluid from the upper mantle drops into the lower mantle, and hot fluid from the lower mantle floods the upper mantle, suddenly raising its temperature. Layering is quickly reestablished and the stratified temperature structure is well established by 1500 Myr (Figure 9.10). After Davies [156]. Copyright Elsevier Science. Reprinted with permission.

indicative. Any thick accumulations of mafic material could be expected to founder once they reached thicknesses of around 60 km and their bases transformed to dense eclogite, in which case the mafic crust would recycle into the mantle anyway.

Between overturns the oceanic crust thickness is only 2–4 km, so its buoyancy would offer little hindrance to subduction. Thus a form of upper-mantle convection, as implicit in the model, is plausible, and these plates would cool the upper mantle. The lower mantle, lacking any contact with the cold surface of the Earth, would lose heat only slowly, and its internal radioactivity actually warms it between overturns. This can be inferred, for example, by the fact that the mean temperature increases as a new overturn is approached. The reason it increases is that heat loss through the surface drops rapidly as the upper mantle cools and drops below the radioactive heating rate (see Figure 9.11(b)).

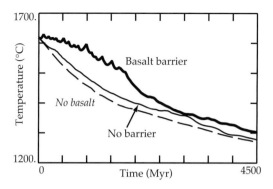

Figure 9.13. Comparison of the mean temperatures of the numerical models of this chapter.

The overturns cease after about 1.8 Gyr in this model (about 2.7 Gyr ago). In other models they ceased after 1.0 and 1.6 Gyr [156]. The reason they cease is that eventually the decline in radioactive heating allows plates to be slow enough and thick enough to penetrate the basalt barrier regularly, thus preventing it from becoming strong enough to block flow. The transition is quite marked (Figure 9.11(a) and (b)), and once it happens the evolution soon converges on a whole-mantle trajectory similar to previous models.

While layering persists, the overall heat loss is inhibited, as just noted – between overturns the upper mantle cools rapidly and its plates slow, so that heat loss is relatively low (Figure 9.11(b)). As a result, the mean temperature of the system stays quite high. This is clearly evident in Figure 9.13, which compares the evolution of the mean temperature in the numerical models of this chapter. Even after the layering has ceased, the model of Figure 9.10 still has a higher mean temperature than the others. This is because there is some inhibition in vertical flow through the transition zone, due both to the vestiges of the basalt barrier (Figure 9.10, lower right panel) and to the assumption in this model that the thermal displacement of phase transformations also inhibits flow through the transition zone, though not so much as to cause layering. (This mechanism has not been dealt with in this book. It is covered in *Dynamic Earth* [1].)

9.4.4 Continental collision

Continental collision is another source of behaviour more complicated than the monotonic declines of Sections 9.1 and 9.2. It is a likely agent of episodicity [157], though the episodicity is more likely to be regional than global. It is included under this heading because it is due ultimately to the compositional buoyancy of continental crust.

The Grenville province of North America is a well-recognised late Proterozoic example of continental collision, and is characterised by high metamorphic grades indicative of deep burial followed by erosional exhumation [158]. Collisions must certainly have contributed to the episodicity of the record, but how much is not so obvious for earlier times. For example, the greatest concentration of ages is in late Archaean (around 2.7 Ga), and the episode seems to have extended at least over a large region, if not globally. However, the terrains from this time are mostly relatively low grade and much new crust seems to have formed, and neither of these features is characteristic of collisions.

9.5 Intermittent plate tectonics

A final potential source of complication is that plate tectonics may have ceased for significant periods [147, 159]. In this case the episodic behaviour might be due to complex rheology rather than to compositional buoyancy.

The re-establishment of subduction after a hiatus might have generated substantial tectonic and magmatic signatures if mantle heat had accumulated significantly during a quiescence. Models have been developed that exhibit something like this kind of behaviour, and an example is shown in Figure 9.14. In this case the episode is due to the accumulation of negative thermal buoyancy in the top boundary layer combined with a nonlinear rheology, which allows strain to become concentrated and the lithosphere eventually to 'break'.

9.6 Implications for tectonics

The goal of this book is to bring readers to the point where they can bring a well-informed general understanding of mantle convection, and mantle convection models, to their own specialty, and thus partake more usefully in continuing discussions and debates. Therefore the implications of the above modelling will not be pursued in great detail, but some general directions of current thought will be briefly surveyed. Thus, for example, there is no attempt here to review comprehensively the substantial literature relating model studies with many kinds of observation.

The near-simultaneous recognition of plate tectonics and mantle plumes provided geology for the first time with well-defined fundamental driving mechanisms for tectonics and the many subsidiary geological processes that accompany tectonic activity. There was, indeed, a revolution in our understanding of the geological Earth. However, it is evident from the geological record that the Earth has not always behaved the way it has for the past 500 million years, or perhaps one or two billion years. This is evident even from a glance at a satellite image of the 3.5 Ga Pilbara block in northwest Australia (Figure 9.15), where the

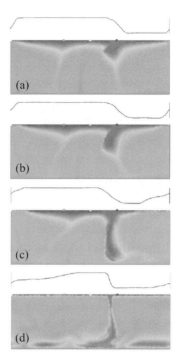

Figure 9.14. A model of an episode of subduction, punctuating a period of static or non-subducting lithosphere. From O'Neill *et al.* [159]. Copyright Elsevier Science. Reprinted with permission.

Figure 9.15. Satellite image of the Pilbara region of northwest Australia. The light blobs are granitoid bodies roughly 50 km across.

9.6 Implications for tectonics

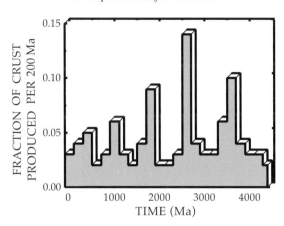

Figure 9.16. A generalised impression of the age distribution of the continental crust. From McCulloch and Bennett [162]. Copyright Elsevier Science. Reprinted with permission.

equi-dimensional granite–greenstone structure bears little resemblance to the elongate, laterally compressed mobile belts typical of later history. There are of course also many more subtle differences between the Archaean and younger geology [160, 161].

There is also a decided clustering of ages of continental rocks and provinces, illustrated in Figure 9.16. This clustering could in principle be due either to episodic tectonic activity or to uneven preservation of continental provinces. The weight of current opinion seems to incline to episodic tectonics [160].

Thus we know that the Earth has changed in important ways, and we know from the preceding models that mantle dynamics has very likely changed in important ways. While some of the modelling results are suggestive, they are far from providing clear interpretations of geological observations. On the other hand, they are suggestive enough to deserve to be included in the multi-stranded conversation that will be required to make sense of the sparse evidence surviving from the Earth's remote past.

Perhaps the primary questions remaining about the Earth's geological history are these: Has plate tectonics always operated? If it has not, what tectonic mode or modes acted in its place? And what role have plumes played? The preceding models have also identified other potential tectonic agents: the mantle overturn following the breakdown of mantle layering, and the cessation and reactivation of subduction and plate tectonics. Continental collisions have been recognised agents independent of modelling. These agents will be discussed in turn.

The results presented in Figures 9.8 and 9.11 show that oceanic-type crust may have been no thicker than at present, despite higher mantle temperatures in the past, because of dynamical depletion of the upper mantle of its more fusible components.

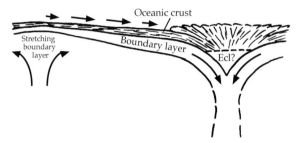

Figure 9.17. Sketch of conjectured 'drip tectonics', in which the top thermal boundary layer of the mantle is not rigid enough to form plates. The structure of downwellings might then not be controlled by brittle faults, so they would tend to be more symmetric. Mafic crust would tend to pile up over downwellings, and if its base became sufficiently deep and warm, it might convert to dense eclogite, triggering foundering of the crust. Such foundering might be highly episodic. From Davies [154]. Copyright Geological Society of America.

This would overcome one potential obstacle to plate tectonics, that the buoyancy of oceanic crust would prevent or slow subduction (Section 9.4.2).

There is no consensus at all as to when plate tectonics started operating. At a conference in 2006, opinions ranged from one billion years ago to 4.5 billion years ago, with a preponderance in the 2–3.5 Gyr range [163]. There is evidence that the Earth's surface was cool enough to sustain liquid water 4.4–4.5 Gyr ago [164, 165] and there is evidence for the presence of some continental crust [165, 166]. This evidence would be consistent with the operation of plate tectonics, but does not require it.

There is evidence for ancient compressional zones that resemble modern subduction zones in many respects, structurally, petrologically and geochemically, though there are also some significant differences [161, 163]. Such zones are known from the mid-Archaean through the Proterozoic. However, such zones do not necessarily imply plate tectonics. The distinctive features of modern subduction zones are that they are structurally asymmetric (only one plate goes down) and that they are narrow mobile belts between large stable blocks. Both distinctions are important for inferring the dynamics of the thermal boundary layer and the mantle. The second distinction might be of little direct concern to geologists, but it ought to be of indirect concern, because the style of mantle dynamics implied could be rather different from the present form, and that would have other important implications, as the results already presented might suggest.

A symmetric compression zone might indicate something like the 'drip tectonics' conjectured by Davies [154] and illustrated in Figure 9.17. In this scenario it is envisaged that the thermal boundary layer is thin enough that it does not act as a rigid unit over very large distances. Thus there might not be anything we would call a plate, and the term *lithosphere*, meaning strong layer, might not apply either. If some of the thermal boundary layer can founder, then it can drive a mode of mantle convection. There could then be a mobile upper thermal boundary layer and therefore fairly efficient removal of mantle heat. If the mafic crust (formed at the extensional zone over a mantle upwelling) were thin, as the models here have suggested, it might simply be pulled down semi-continuously by boundary layer drips. On the other hand, if the mafic crust were thick enough, it might resist foundering and accumulate into thick piles. If such piles reached a thickness of about 60 km or greater, the lower parts might transform to dense eclogite. These would then tend to sink fairly rapidly, and they would pull shallower material after them. Thus a mafic 'sinker' might founder into the mantle, and this might occur in a very episodic manner.

It is not at all clear that such a regime ever operated on Earth, although it is one way in which the Archaean nuclei, like the Pilbara in Figure 9.15, might have formed. This scenario nevertheless serves to expand thinking about possibilities, and to clarify what needs to be established before it can be concluded that plate tectonics operated. Many of the structural features of such a compression zone might resemble a modern subduction zone. Many of the petrological and geochemical signatures might also resemble modern subduction zones, because the primary petrological process in both scenarios may be the remelting of hydrated mafic material at significant depth. One way in which petrology and geochemistry might provide key evidence is if they revealed asymmetry of the overall structure, the way the magmatic arcs do in modern subduction zones.

Apart from asymmetry, another key criterion for plate tectonics is the presence of plates outside the active belts. This may be very difficult to demonstrate, because of course there is little evidence generated outside of mobile belts.

To summarise this brief discussion of whether plate tectonics or some other regime operated earlier in Earth history, it is plausible that sufficient evidence will be accumulated to demonstrate the existence of something fairly close to a modern asymmetric subduction zone. The asymmetry would imply a lithosphere acting more as a brittle solid than as a malleable fluid, and it would suggest that plates may be present. However, it would not directly demonstrate the presence of such plates outside the mobile belt, nor indicate how large they might have been.

Turning to plumes, arguments were given in Chapter 7 that plumes are to be expected in the silicate mantles of Earth-like planets. The models shown here indicate that plumes may have been at least as active as they are now. Some versions of such thermal history calculations can yield considerably higher activity in the past than now [65]. Plumes are secondary tectonic agents and their record may be more ephemeral than that of subduction zones, both because much of

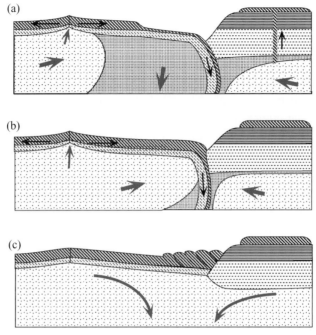

Figure 9.18. Possible tectonic events accompanying a mantle overturn. Hot material from the lower mantle (stippled) replaces cooler material in the upper mantle (grey). As a result, thicker oceanic crust is formed, and flood-basalt-like eruptions occur on continents. The thickened oceanic crust blocks subduction in (c). From Davies [175]. Copyright Elsevier Science. Reprinted with permission.

their record is left in ocean basins and because their basaltic products are readily weathered away.

One kind of candidate for inferring past plume activity may be large igneous provinces, inferred to come from mantle plumes [170]. Indeed the possibility that Archaean greenstone belts are the products of mantle plumes was an important stimulus to developing the modern theory of plume heads [167, 168]. Another signature of a plume head may be radiating dyke swarms, inferred to be the feeder dykes of flood basalt provinces [169, 170].

The apparent episodicity of the continental record requires something beyond the smoothly evolving models of Figures 9.1, 9.2 and 9.6. The basalt barrier mechanism of Section 9.4.3 certainly induces episodic behaviour of the right general kind, especially in having overturns that would produce major magmatic and tectonic episodes. The material rising from the lower mantle is not only 200–300 °C hotter than the upper mantle, but also more fertile. The volume of magma potentially involved, as indicated by the spikes in the thickness of oceanic crust, is perhaps 200 times more than in a flood basalt eruption. Figure 9.18 sketches how such an

9.6 Implications for tectonics

event might develop. Whether Archaean greenstone belts are the product of plume heads or overturns remains to be resolved.

The tectonic regime responsible for the earliest Archaean nuclei is inferred to be more 'vertical' in character than plate tectonics [171, 172]. Reheating of a pre-existing mafic crust is inferred, perhaps in repeated episodes. The agents of such reheating might be very large plume heads or mantle overturns. A significant point emphasised by Bédard [171] is that the persistence of thick lithospheric keels associated with the Archaean nuclei indicates that we should think of the keel and the crust forming in intimate association, and in that sense the keel would be a differentiated product of the mantle just as the crust is.

There is active work on attempts to document past continental aggregations [173, 174], and one of several tools is the identification of past continental collisions. The whole cycle of continental dispersal and re-aggregation, often called the Wilson cycle, is likely to contribute to the episodic record. However, as noted earlier, the episodes are more likely to be regional than global, and so do not have great implications for the operation of mantle convection. Nor are studies of mantle convection so likely to contribute to their understanding. Rather, the meanderings of continents may be a second-order modulation of the plate–mantle system.

On the other hand, if the plates ceased to move for significant periods of Earth history, that would affect the evolution of the mantle and require a deeper understanding of the mechanics of the plate–mantle system. If there was a hiatus in plate activity, then the mantle heat loss would fairly quickly decay to fairly low levels and the mantle temperature might well begin to increase. As noted earlier, so long as the hiatus did not last longer than a few hundred million years, there might not be a large effect on overall geological activity. However, the re-establishment of subduction and plate tectonics would probably initiate an episode of fairly high activity, as the higher mantle temperature accumulated in the hiatus would have lowered mantle viscosity and would thus increase plate speeds. The higher temperature would also result in greater degrees of melting. Thus a cycle of hiatus and renewed activity could contribute significantly to the episodic nature of the geological record. A hiatus would also help to resolve the heat source puzzle discussed in Section 9.3.

Intermittent plate tectonics would bring to the fore the complex rheology that undoubtedly characterises rocks in the varied pressure–temperature regime involved in mantle convection. Much of our understanding of mantle convection has been built on the assumption that the complications mostly do not matter very much, but intermittency would immediately imply that the complications must be given greater consideration, however difficult that might make modelling studies.

10

Mantle chemical evolution

> Trace element heterogeneity of the mantle; apparent ages. Global budgets and a mildly depleted MORB source. Distinct OIB source, no primitive mantle. Major element heterogeneity, sources and survival. Melting reconsidered; physical partitioning, disequilibrium, melt trapping. Recycling oceanic crust and hybrid pyroxenites. Critique of previous abundance estimates. Geochemical modelling using tracers in convection models. Density differences. Residence times. Ages due to remelting, not homogenising. Incorporating noble gases.

The chemistry of the mantle has already entered this presentation implicitly and explicitly. It is explicit in the discussions of compositional differences and radioactive heating, and it is implicit in that the material properties we have called upon depend of course on the composition of the relevant materials. The geochemistry of the mantle is divided for convenience into major elements, trace elements and isotopes. Trace elements and their isotopes give us some key information, but in order to interpret them properly we need to consider the major elements as well.

Trace elements and isotopes give us some important, though indirect, information about the structure of the mantle, and the isotopes give us some time information as well. These are very important kinds of information that are not available from other sources, so we should take advantage of them if we possibly can. It is turning out that the interpretation of these geochemical observations requires a careful consideration of both geochemical and geophysical processes. This should not be too surprising, but there have been many proposed interpretations using only one side or the other of the available information, with the result that two quite incompatible pictures of the mantle were built up, one layered and the other not. This incompatibility of interpretations obviously has implied that there were important things we did not understand. The early tendency, in seeking resolution, was simply to assert that the other side must be wrong, but over the past 15 years or so a generally more constructive conversation has developed.

The main lessons from trace elements and isotopes can be summarised fairly simply. There are detectable heterogeneities in trace element concentrations, and in isotopic ratios, inferred to exist in the sources of rocks derived from the mantle. Variations in lead isotopes indicate that the heterogeneities are old, roughly 1.8 Ga. Some quite volatile elements, and in particular the noble gases, are still leaking out of the mantle in detectable quantities. Rocks from volcanic hotspots, termed oceanic island basalts (OIBs) by geochemists, have greater concentrations of 'incompatible' trace elements and show greater isotopic variation than mid-ocean ridge basalts (MORBs). ('Incompatible' means the element is not well accommodated by mantle minerals, and it tends to be extracted preferentially by erupted mantle melts.) About half of some of the most incompatible trace elements have been sequestered into continental crust and the atmosphere. Data supporting these propositions are presented, initially with little interpretation.

The interpretation of the geochemical observations requires an appreciation that the mantle is heterogeneous also in major elements, and of the implications of that heterogeneity for how the mantle melts and how that melt migrates, or does not migrate, through the heterogeneous mantle. Such appreciation has grown only slowly, and presentations hitherto generally still have included assumptions based on melting a homogeneous source. Those assumptions need to be re-evaluated before they can be the basis of reliable interpretations. Therefore the interpretations to follow are presented differently in an effort to avoid such unjustified assumptions.

10.1 Trace element and isotope observations

The data on which the above inferences are based are being added to all the time, but the main patterns have been evident for some time, so some older illustrations will suffice, though more up-to-date versions of some can be found, for example, in Hofmann [60].

Figure 10.1 shows patterns of trace element concentrations that illustrate the differences between MORBs and OIBs. The elements are ordered, left to right, from most incompatible to most compatible. MORBs have much lower concentrations of highly incompatible elements than OIBs, though Hawaii is intermediate. In this plot OIBs are fairly similar to continental crust, though they differ substantially in many other respects. Nb, Pb and Ti are anomalous in several of the trends.

The primitive mantle of Figure 10.1 is the estimated composition prior to the extraction of the continental crust [140]. It is, in other words, the average composition of the silicate parts of the Earth, crust plus mantle. All of the

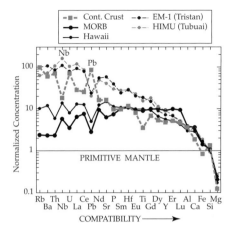

Figure 10.1. Trace element concentrations in mantle-derived rocks. Concentrations are normalised to the estimated concentrations in the primitive mantle. The Hawaii, Tristan and Tubuai samples are ocean island basalts. EM1 (EM = enriched mantle) and HIMU (high mu) are different classes of OIB. After Hofmann [176]. Reprinted from *Nature* with permission. Copyright Macmillan Magazines Ltd.

incompatible elements in both OIBs and MORBs have concentrations higher than primitive. For MORBs this is inferred to be due to their being concentrated into melts during melting, as will be discussed later.

Differences between OIBs and MORBs are evident in isotopic ratios as well. Figure 10.2 summarises isotopic ratios for strontium, neodymium and lead. MORBs show less variation than OIBs (Pacific MORB is the most visible in the plots), although Indian Ocean MORB is intermediate in variation. Some of the variation may be due to the incorporation of material from continental crust, as is suggested in panel (c), but much of the variation is intrinsic to the mantle/oceanic crust system. The Sr–Pb plot (panel (a)) makes clear that at least three components or source types would be required to explain the range of variation, and in fact five or six have been proposed [177, 178]. Most of the variations evident in OIBs are also evident in MORBs, but with a smaller range of variation.

The uranium–lead radioactive decay system is especially useful because it contains two radiogenic isotopes, and in combination they give age information [179]. In Figure 10.3(a), the slope of the overall correlation corresponds to an age of about 1.8 Ga. This would only be an actual age if all samples originated from the same homogenised source at the same time, which they obviously will not have. However, it does indicate an approximate aggregate age, though probably with some bias to greater age. The significance of the correlation will be clarified later. In the meantime, the correlation does indicate that the isotopic heterogeneities evident in Figures 10.2 and 10.3 have survived in the mantle for the order of one to two billion years.

10.1 Trace element and isotope observations

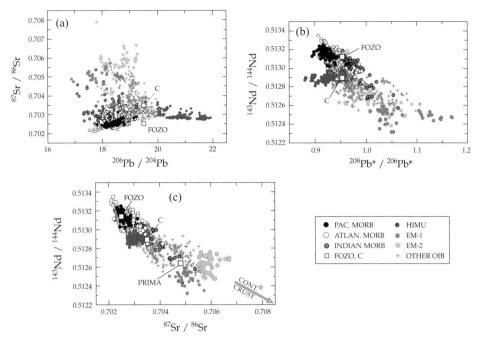

Figure 10.2. Isotopic data from oceanic mantle-derived rocks. Each ratio is of a radiogenic isotope over a non-radiogenic reference isotope. EM2 is another class of OIB. FOZO and C are proposed mean compositions. PRIMA is estimated primitive mantle. ^{208}Pb* and ^{206}Pb* refer to just the radiogenic components of these isotopes. After Hofmann [176]. Reprinted from *Nature* with permission. Copyright Macmillan Magazines Ltd.

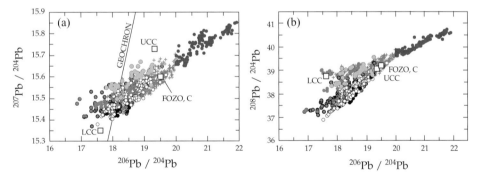

Figure 10.3. Lead isotopic data for oceanic mantle-derived rocks. ^{207}Pb is the radiogenic daughter of ^{235}U, ^{206}Pb is from ^{238}U, and ^{208}Pb is from ^{232}Th. UCC and LCC refer to the upper and lower continental crust. The *geochron* is a 4.5 Ga isochron tied to an iron meteorite. After Hofmann [176]. Reprinted from *Nature* with permission. Copyright Macmillan Magazines Ltd.

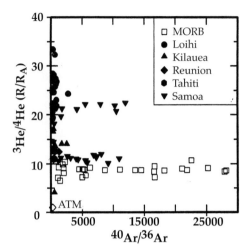

Figure 10.4. Helium and argon isotopes from MORBs and from various ocean island groups. The helium data are normalised to the atmospheric ratio ($R_A = 1.4 \times 10^{-6}$, [180]), denoted by the point ATM ($^{40}Ar/^{36}Ar \sim 300$). After Porcelli and Wasserburg [181]. Copyright Elsevier Science. Reprinted with permission.

Important information comes from observations of helium, neon and argon in mantle-derived rocks. These gases occur in minute concentrations and the concentrations are subject to near-surface processes, so concentrations do not give a very robust basis for interpretation. Therefore most attention is focused on isotopic ratios. Figure 10.4 summarises observations of helium and argon. ^{4}He is produced by radioactive decay of U and Th, and ^{40}Ar is produced by decay of ^{40}K. Unfortunately, helium is usually presented with the reference isotope, ^{3}He, in the numerator, $R = {}^3He/{}^4He$, contrary to the convention for other elements.

Figure 10.4 shows that argon ratios in the mantle are generally much more radiogenic than those in the atmosphere, whereas the helium ratios are generally much less radiogenic than those in the atmosphere. For argon, the main reason for this is that argon accumulates in the atmosphere and comprises about 1% of the atmosphere, so the atmospheric abundance of ^{36}Ar is relatively high. On the other hand helium continuously escapes from the top of the atmosphere so that helium is a minor part of the atmosphere. The abundance of U and Th in the continental crust means that the crust contributes substantial ^{4}He to the atmosphere, making it more radiogenic.

The helium ratios for MORBs are relatively uniform, around 8±1. In contrast the OIB helium is quite variable, ranging from about 4 to over 30. Thus some OIB helium is less radiogenic than MORB helium (i.e. above 8) and some is more radiogenic.

10.1 Trace element and isotope observations

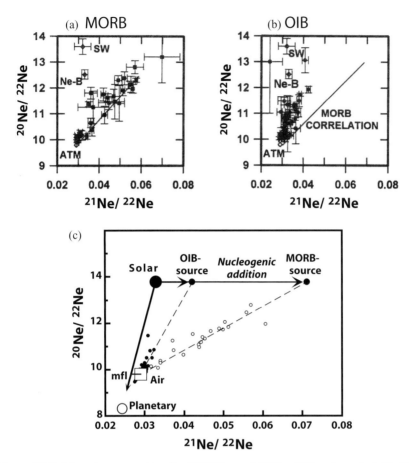

Figure 10.5. Neon isotope ratios for MORBs and OIBs. SW: solar wind. Ne-B: 'neon-B' component in meteorites. (a) MORB data, (b) OIB data, (c) interpretation. ATM: atmosphere. mfl: mass-fractionation line. (a) and (b) Modified from Porcelli and Wasserburg [181]. Copyright Elsevier Science. Reprinted with permission. (c) After McDougall and Honda [180].

The argon data show that both MORB and OIB ratios are quite variable, but the MORB data range to higher values than OIB data. The highly radiogenic values of both types are inferred to reflect the fact that argon is very incompatible in the mantle and most of it has escaped, leaving very low concentrations of ^{36}Ar. On the other hand, potassium is relatively compatible, so its abundance has probably not changed by more than a factor of 2 or so and ^{40}Ar keeps accumulating in the mantle at a significant rate.

Neon observations are summarised in Figure 10.5. Three isotopes are observed. ^{21}Ne is produced in the Earth by the reactions of neutrons, produced mainly by the decay of U and Th, with other elements. The other two isotopes are not nucleogenic.

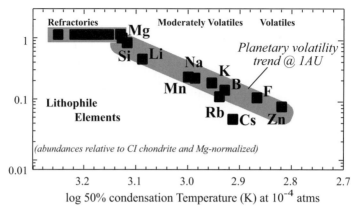

Figure 10.6. Abundances of lithophile elements as a function of condensation temperature in the solar nebula, normalised to chondritic meteorite abundances. The more volatile elements are systematically depleted. From McDonough [183]. Copyright Elsevier Science. Reprinted with permission.

Neon is inferred to have started in the mantle with a composition similar to solar wind (SW) or meteoritic (Ne-B) neon. Nucleogenic ^{21}Ne is assumed to have been progressively added (Figure 10.5(c)). The arrays of data evident in Figures 10.5(a) and (b) are inferred to result from contamination with atmospheric neon during eruption at the Earth's surface. The atmospheric composition is believed to have resulted from mass fractionation during early loss from the top of the atmosphere, which would move it along the 'mfl' line, followed by a small addition of ^{21}Ne from the mantle.

It is clear that OIBs tend to fall along a less nucleogenic trend than MORBs (Figures 10.5(a) and (b)). The mantle sources are inferred to lie at the intersections of these trends, with the mantle line evolving by nucleogenic addition (Figure 10.5(c)). Thus the OIB sources are inferred to be less nucleogenic than the MORB source. These relationships are similar to those of helium isotopes, reflecting the fact that additions come from U and Th for both elements.

10.2 Global budgets

Studies of meteorites have been used to estimate the total amounts of some elements that were incorporated into the Earth [140, 182]. Elements that are refractory in the environment of the early solar nebula are believed to have been added to the Earth in unmodified relative proportions, whereas more volatile elements were blown out of the Earth's formation zone by the solar wind, so that the more volatile elements are systematically depleted within the Earth. Figure 10.6 shows estimates of normalised abundances of lithophile elements (i.e. those that prefer the silicate

Table 10.1. *Uranium and heat generation budgets*[a].

Reservoir	Mass (10^{22} kg)	Mass (%)	U conc. (ng/g)	U mass (10^{15} kg)	Heat gen.[b] (pW/kg)	Heat gen. (TW)
Average silicate Earth	400	100	20[c]	80	5	20
Continental crust	2.6	0.65	1400[d] (900–1800)	36 (23–47)	350 (225–450)	9 (6–12)
D″	8.5	2.1	50[e] (50–80)	4 (4–7)	10.5 (10–15)	1
MORB source, i.e. rest of mantle	389	97	**10** (6.5–13)	**40** (26–53)	2.5 (1.5–3)	10 (7–13)

[a] (Plausible ranges in parentheses.) **Bold** numbers inferred from U mass balance.
1 pW = 1 picowatt = 10^{-12} W. 1 TW = 1 terawatt = 10^{12} W.
[b] Heat generation productivity from Stacey [141], assuming Th/U = 3.8 and K/U = 13 000.
[c] McDonough and Sun [140], O'Neill and Palme [182] and Lyubetskaya and Korenaga [187].
[d] Rudnick and Fountain [142].
[e] Assumed similar to oceanic crust; Sun and McDonough [188] and Donnelly *et al.* [189].

crust and mantle rather than the metallic core), and the trend to lower abundance with higher volatility is evident.

It is useful to follow the arguments for uranium, as it is one of the refractory elements whose total abundance in the Earth is reasonably well known, it is one of the incompatible elements that is partitioned strongly into melt phases in the mantle, and it is a heat source element. Table 10.1 summarises estimates of uranium concentration in the Earth and in its main reservoirs, as they have been inferred in preceding chapters. The average U content of the primitive mantle is inferred from meteorites to be 20±4 ng/g, which implies a total mass of 80×10^{15} kg of U. About $(23–47) \times 10^{15}$ kg is estimated to be in the continental crust, i.e. 30–60%. Thus $(33–57) \times 10^{15}$ kg or 40–70% is inferred to be in the mantle.

Some of this mantle uranium we can plausibly locate in the seismic D″ zone at the bottom of the mantle. D″ is interpreted here as including the accumulations of subducted oceanic crust hypothesised by Hofmann and White [74] and demonstrated by models like those of Figures 9.5 and 9.10. The accumulation is taken, from the numerical models, to be equivalent to a layer about 100 km thick [118]. If its U content is similar to that of oceanic crust, then it would contain only about 5% of the total U budget.

According to the mantle picture inferred in previous chapters, the only remaining mantle reservoir is the rest of the mantle, in other words excluding the crust and the D″ layer. The balance of the Earth's uranium is then required to be in this reservoir,

162   *Mantle chemical evolution*

and Table 10.1 shows that it would have a concentration of about 10 ± 3 ng/g. This is significantly higher than traditional estimates of 3–4.7 ng/g [184–186]. The reasons for the previous answers will be discussed later, though we can note from Figure 9.8 that abundances estimated from MORB concentrations may be too low by a factor of up to 2 or so, because of the gradient of basalt tracers through the upper mantle. Similar mass balances apply to other refractory incompatible elements.

10.3 Incompatible trace elements in the mantle

We are now in a position to interpret some features of Figure 10.1, starting with the enrichment of MORBs in incompatible trace elements, again using uranium as an example. MORB, at about 75 ng/g in the figure, is enriched relative to both the primitive mantle (20 ng/g) and our inferred MORB-source mantle (10 ng/g). The enrichment of MORB relative to its source is interpreted as resulting from the concentration of U into the melt phase during melting, because of the incompatibility of U in the crystal structures of mantle minerals. The concentration in the present interpretation is by a factor of 7.5 (or a range of 5–8).

At this point in conventional interpretations, this concentration factor would be used to infer that the MORB source underwent about 13% partial melting, but we cannot make this inference if the source is heterogeneous. This will be discussed in Section 10.4. Actually, in the conventional interpretation, the concentration factor is 16–25 (75 ng/g relative to 3–4.7 ng/g) and the inferred degree of melting would be 4–6%, or 6–10% if MORB is taken to average 50 ng/g.

The continental crust is enriched in the most incompatible elements by a factor of 80–100 relative to primitive (Figure 10.1). It was recognised early that the depletion of the MORB source is likely to be complementary to the enrichment of the continental crust [176]. This interpretation is implicit in the mass balance of Table 10.1. However, previous mass balances, based on what are argued here to be underestimates of the MORB-source abundances, indicated that only 30–60% of the mantle had been so depleted, and the rest was assumed to be primitive. It will be shown in the next section that we cannot expect a significant amount of primitive mantle to have survived. Nevertheless the basic complementarity of the mantle and the continental crust is a straightforward inference from the identified reservoirs in the Earth, as was illustrated in compiling Table 10.1.

Hofmann [190] demonstrated that the general pattern of the continental crust in Figure 10.1 could be accounted for by about 1% partial melting of a primitive mantle (see also [60]). He stressed that this is clearly not the process by which the continental crust was formed, but the effect of the relative compatibilities persists through the real process, which involves multiple stages. This conclusion will remain true in the present interpretation.

10.3 Incompatible trace elements in the mantle

The OIBs in Figure 10.1, apart from Hawaii, have enrichment patterns comparable to that of continental crust. A naive inference would be that OIBs also result from about 1% partial melting of the mantle, but Hofmann and White [74] argued that the degree of melting of OIB sources was larger (of the order of 5%), and that the OIB sources must be enriched in incompatible trace elements relative to MORB sources. Their interpretation is supported by the fact that Iceland shows enrichments, even though it is located on a spreading centre and the degree of melting is likely to be as great or greater than for MORBs.

The relatively low enrichment of Hawaii is interpreted as due in part to a relatively high degree of melting, which tends to dilute the concentrations of incompatible elements. This is consistent with Hawaii being much the strongest plume, as we saw in Chapter 7, which would make it less prone to lose heat ascending through the mantle, and thus more likely to melt more. However, the low enrichment may also reflect a less enriched source than those of other OIBs.

Hofmann and White went on to propose that the enrichment of the OIB sources is due to the incorporation of a higher proportion of old subducted oceanic crust than in MORB sources. They proposed that subducted oceanic crust accumulated at the base of the mantle, in the D'' zone, and the plumes entrain some of this crust and carry it to the top of the mantle, where it melts preferentially. This interpretation is supported by the patterns of anomalous elements in Figure 10.1. Nb is low and Pb is high relative to the general trend of continental crust. MORBs show the reverse pattern, which is consistent with the MORB source being the complement to continental crust. The OIB anomalies resemble the MORB anomalies, which fits with them being due to an excess of old MORB. If OIBs were due just to low-degree melting of normal mantle, the Nb and Pb anomalies might be expected to resemble those of continental crust, or to have no anomaly at all.

This relationship of OIBs with MORBs was made more explicit by Hofmann *et al.* [191], who displayed the Nb/U ratio for MORBs and OIBs relative to primitive mantle and continental crust. A later version of their plot is shown in Figure 10.7, and a more sophisticated argument, reaching the same conclusion, is given by Hofmann [60]. The plot shows that the OIBs resemble the MORBs in mostly having a Nb/U ratio higher than primitive, rather than resembling continental crust, which has a lower ratio. A few of the EM2 OIBs have ratios comparable to primitive, but these can be interpreted as containing a few per cent of continent-derived sediment.

Figure 10.7 also argues against significant involvement of primitive mantle in OIBs. Early views of the Nd isotopes of Figure 10.2(c) were that the spread of data is due to mixing between a highly depleted component and a primitive component. However, most OIBs have a distinctly higher Nb/U ratio than primitive, and the few that are similar are better explained by mixing of continental components than by primitive components.

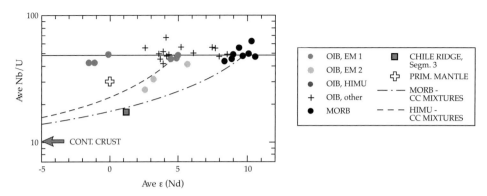

Figure 10.7. Average Nb/U versus Nd isotopic composition for suites of mantle-derived oceanic rocks. ε(Nd) is the deviation, in parts per 10 000, from the primitive mantle ^{143}Nd/^{144}Nd. From Hofmann [176]. Reprinted from *Nature* with permission. Copyright Macmillan Magazines Ltd.

10.4 Mantle heterogeneity

The mantle is chemically heterogeneous on all scales, from mineral grains through hand specimens and melting zones to ocean basins. This ranks as a primary observation, and it is central to the interpretation that follows. Although this heterogeneity has become more widely recognised [60], its implications have not been fully explored. Furthermore, the heterogeneity is just as evident and important for MORBs as for OIBs, though OIBs span a larger range of values, typically about double the range of MORBs, depending on the isotopic system.

Heterogeneity of MORBs at the largest scale is evident in Figure 10.2, where differences can be seen particularly between the Indian Ocean and the Pacific Ocean, with the Atlantic Ocean tending to be intermediate. These differences are even more clearly illustrated in colour by Hofmann [60]. Heterogeneity at intermediate scales, down to a few degrees of latitude, are evident in Figure 10.8. It is evident that there are variations at the scale of a few degrees comparable to the variation along the entire ridge, for example near 14 °N. Large variations have also been found within single hand specimens [60], suggesting that such variations are present in the source, though they will tend to be homogenised within magma chambers during extraction.

The heterogeneity is evident not just in trace elements but also in major elements. This is inferred from variations in rocks derived by melting the mantle [192–194]. It is also inferred from occasional outcrops of mantle material [195]. Mantle rocks are commonly observed to comprise two main rock types, the dominant peridotite and secondary, but common, eclogite [196]. Some care is required in attributing these observations, because some of the detail in those rocks may reflect the particularities

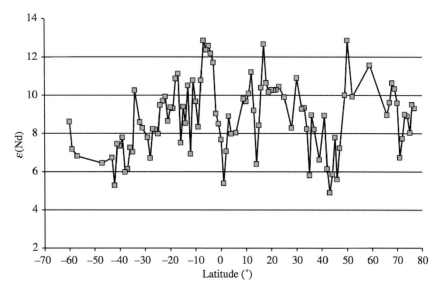

Figure 10.8. Neodymium isotopic composition versus latitude (degrees north) along the Mid-Atlantic Ridge. Data have been averaged over 1° intervals. ε(Nd) is the variation in ^{143}Nd/^{144}Nd, measured in parts per 10 000. From Hofmann [60]. Copyright Elsevier Science. Reprinted with permission.

of their emplacement as much as their original condition. Nevertheless, few seem to doubt that the mantle is indeed a heterogeneous mixture of rock types.

10.4.1 Sources of heterogeneity

There is an obvious source of heterogeneity, in the subduction of differentiated oceanic lithosphere. The differentiation occurs at mid-ocean ridges, where melting forms the oceanic crust and leaves a depleted zone, perhaps 60 km thick, from which the melts have been extracted. Hydrothermal alteration of the near-surface crust, perhaps down several kilometres, and the deposition of pelagic sediments add additional chemical heterogeneity. Some of this will be removed by melting during subduction, but the major heterogeneity will persist.

An impression of the subsequent dispersal of subducted lithosphere is provided by Figures 9.5 and 9.10. The heterogeneities tend to be reduced down to smaller scales with time, but the figures show that the heterogeneities persist visibly throughout the fluid even after the equivalent of billions of years. This will be true in the real mantle as well, even as heterogeneities are reduced to scales of kilometres or less, because solid-state diffusion rates are so slow that chemical homogenisation would only occur over scales less than about one metre even over billions of years [197].

It turns out that plumes are also a significant source of heterogeneities. The mass flow of plumes is less than that of subducted lithosphere, but only by a factor of about 3. The rate of subduction of sea floor is $3\,km^2/yr$ (Chapter 6). The strongly differentiated part of the lithosphere is about 60 km deep, so the volume of heterogeneous lithosphere added per year to the MORB source is around $180\,km^3/yr$. On the other hand, the volumetric flow rate of the Hawaiian plume is estimated to be about $7.5\,km^3/yr$ (Chapter 7) and the Hawaiian plume carries about 10% of the global plume flow, so the global volumetric flow rate of plumes into the upper mantle is about $75\,km^3/yr$, around 40% of the plate flow, or 30% of the combined flows into the MORB source.

Most of the plume heterogeneity will not be removed by melting, because only 10–20% of the plume material actually melts (Chapter 7), and the rest will be stirred into the mantle [1]. Thus we should expect around one-quarter of mantle heterogeneity to derive from plumes. These heterogeneities may ultimately derive from subducted lithosphere, but they may also be modified in important ways prior to their incorporation into a plume, as we will see.

10.4.2 Survival of heterogeneities

The early intuition of many geochemists was that mantle convection would keep the mantle homogenised, and this led many to assume that the observed heterogeneity required distinct layers to preserve chemical differences for long periods. This intuition was probably fed by experience with fluid mixing in daily life, where, for example, milk added to coffee can be homogenised with a quick stir of a spoon. However, this experience is quite misleading, because the liquid in a coffee cup is in a quite different flow regime from the mantle. The key difference is that water has a sufficiently low viscosity that the momentum of moving fluid sustains its motion and, even more important, generates smaller-scale eddies spontaneously. There is a cascade of flow into smaller and smaller eddies, and the eddies of different scales combine to mix the fluids very efficiently. This regime of flow is called turbulence. In contrast, the mantle has such a high viscosity that momentum is completely negligible and there is no turbulence. This means that the smaller-scale eddies that so efficiently homogenise coffee are absent in the mantle. Homogenisation can occur only by the slow shearing and stretching caused by the large-scale flow, and this takes many orders of magnitude longer. Try stirring milk into honey and see how much longer it takes. This regime of flow is called laminar flow.

Quite apart from this fundamental distinction, there has been some debate among geophysicists about whether mantle homogenisation might take hundreds of millions of years or (many) billions of years. The rate of stirring depends on details of the flow, such as whether there are flows at scales smaller than the plates, whether

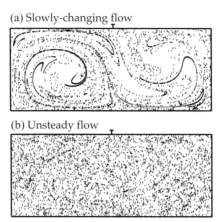

Figure 10.9. Stirring rates in different types of flow. (a) Flow that changes only slowly, in the manner of slowly evolving plates. (b) Flow that changes from one cell to three and back, in the manner of unsteady convection in constant-viscosity fluids. The tracers began in compact clumps, and have been stirred for the equivalent of about 2 Gyr in each case. From Davies [1].

flows change rapidly with time and how important the three-dimensionality of flow is [198]. Some of the early models assumed that convection was confined to the upper mantle and used constant-viscosity fluid with no plates, so the scale of flow was only a few hundred kilometres, flows were quite time-dependent, and homogenisation indeed required only a few hundred million years [199]. However, models with plates yield much larger scales of flow that change only slowly, and mixing can take many billions of years in such flow [200].

Figure 10.9 illustrates one aspect of this, showing that a flow with an unrealistically variable flow homogenises tracers much more quickly than a flow that changes in the manner of slowly evolving plates. Much of this debate was motivated by attempts to explain the age of heterogeneities inferred from lead isotopes (Figure 10.3) as reflecting the homogenisation timescale [201], but we will see later that a different explanation has come to the fore, and the timescale of homogenisation is a less important consideration. In any case the observations of geochemical variations cited earlier give a strong basis for assuming that the mantle is indeed heterogeneous on all scales.

10.4.3 How much primitive mantle?

The process of melting under mid-ocean ridges, to form oceanic crust and its depleted mantle complement, is the main source of the heterogeneities that are later subducted and stirred into the mantle. If this process has been operating for several billion years, how much old oceanic crust would have accumulated in the mantle? And if the process, or something like it, began early in Earth history, then it would have been converting primitive mantle into processed, heterogeneous mantle. Is there likely to be any primitive mantle left? We can address the second question first.

The depth of the main melting under mid-ocean ridges (i.e. the depth to the dry peridotite solidus) is usually taken to be about 60 km, but minor melting or melting of heterogeneities might occur as deep as 110 km [80, 202]. The present areal rate of seafloor spreading is 3 km²/yr. The rate at which mantle mass is being processed through the ridge melting zone can then be calculated as

$$\phi = \rho A_s d_m, \tag{10.1}$$

where ρ is the density of the upper mantle, A_s is the areal spreading rate and d_m is the melting depth. The time it would take to process one mantle mass, M, at this rate is then

$$\tau = M/\phi. \tag{10.2}$$

This can be called the mantle processing time [203]. With the above values, a density of 3300 kg/m³ and a mantle mass of 4×10^{24} kg, this gives $\tau = 4$ Gyr.

This result suggests that most of the mantle will have been processed through ridge melting zones, but there are two other important things to consider. On the one hand, the mantle was probably convecting faster in the past, which would increase the amount of processing. On the other hand, some of the material processed may have been previously processed, and this reprocessing will not affect the amount of primitive mantle remaining.

As we saw in Chapter 9, because of higher radioactive heating in the past, the mantle would have been hotter, would have had a lower viscosity and so would have convected faster. At present the average plate velocity is about $v = 5$ cm/yr. A useful timescale to use here is the *transit time*, which is the time it would take mantle material to traverse the mantle depth, $D = 3000$ km. The transit time is then

$$t_t = D/v, \tag{10.3}$$

and this evaluates to $t_t = 60$ Myr. If convection had been proceeding at the present rate for 4.5 Gyr, there would thus have been time for about 75 transits. From a thermal evolution calculation of the kind presented in Chapter 9 (but neglecting the early transient cooling), I estimated [203] that in fact there have been about 300 transits, four times as many as assuming a steady rate. At present rates of convection, it would thus take 4×4.5 Gyr $= 18$ Gyr to accomplish this number of transits.

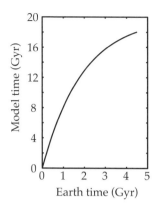

Figure 10.10. Conversion from model time to Earth time. After Davies [203]. Copyright Elsevier Science. Reprinted with permission.

In 2002, because of computer limitations, it was still difficult to achieve the numerical resolution required to run convection models of the early, hot mantle like that presented in Section 9.2. Instead, I ran a model at present conditions and present rates for 18 Gyr, so as to achieve the appropriate number of mantle transits. I then converted the numerical model time, t_m, to the real time of a cooling mantle. The relationship between the model time and real time is shown in Figure 10.10, and a formula is given in Davies [203]. The same idea can be used to estimate the amount of primitive mantle remaining, i.e. we can calculate how much mantle would be processed within 18 Gyr at present rates of processing.

Now we are ready to consider that some of the material melting under a ridge may already have been melted in the past. Suppose the mass of the mantle that is still primitive at a particular time, t, is m. Then the fraction of primitive mantle is $p = m/M$. If, for example, $p = 0.4$, then 40% of the mantle is primitive and 60% has been processed through a melting zone. If the chance of a piece of mantle being processed within the next little while is independent of whether it has already been processed, then, on average, only 40% of the next batch processed will have been primitive.

Going back to the beginning, when $t = 0$ and $p = 1$, the mass of mantle processed within a time Δt will be $\phi \Delta t$, where ϕ is given by Eq. (10.1). Then the mass of primitive mantle will decrease from $m = M$ to $m = (M - \phi \Delta t)$. In other words the *change* in m is $\Delta m = \phi \Delta t$, and the *rate of change* is $\Delta m / \Delta t = \phi$. Later, however, when $p = 0.4$, the change in m is only 40% of this, because 60% of the mass processed has already been processed. In other words, we can write

$$\frac{\Delta m}{\Delta t} = -p\phi$$

and, remembering that $p = m/M$ and dividing both sides by M, we get

$$\frac{\Delta p}{\Delta t} = -p\frac{\phi}{M} = -\frac{p}{\tau}, \tag{10.4}$$

where the second step uses Eq. (10.2). If we just used the initial rate of change of p, when $p = 1$, we would predict that p would be reduced to zero after 4 Gyr, which is the processing time. However, Eq. (10.4) says that the rate of change of p gets smaller as p gets smaller, and therefore p will approach zero asymptotically. Section C.1 in Appendix C shows that the mathematical solution of this equation is that p declines exponentially with time:

$$p = \exp(-t_m/\tau), \tag{10.5}$$

where t_m denotes that we are working in the 'model time' of Figure 10.10.

After $t_m = 18$ Gyr with $\tau = 4$ Gyr, $p = 0.011$. Thus by this estimate only about 1% of the mantle remains primitive. This fraction would be even smaller if the melting depth was greater in the past, because the mantle was hotter. This simple calculation means that we cannot expect a significant amount of primitive mantle to have survived in the MORB source.

10.4.4 How much subducted oceanic crust?

Oceanic crust is being added to the mantle at subduction zones at the rate

$$\phi_c = \rho_c A_s d_c, \tag{10.6}$$

where ρ_c is the density of oceanic crust and d_c is its thickness. We can ask how long it would take, at present rates, to fill the mantle with oceanic crust. The answer is $\tau_c = M/\phi_c$. With a crustal density of 2900 kg/m³ and thickness of 7 km, $\phi_c = 6.1 \times 10^{13}$ kg/yr and $\tau_c = 66$ Gyr. As the Earth is only 4.5 Gyr old, you might conclude that oceanic crust comprises a fraction $f = 4.5/66 = 0.07$ of the mantle, so old subducted oceanic crust would comprise 7% of the mantle. However, if subduction was faster in the past, as we are supposing, you should allow for the equivalent of 18 Gyr worth of subduction at present rates, so $f = 18/66 = 0.27$, and crust would comprise 27% of the mantle, which seems like a lot.

However, oceanic crust is also being removed in the melting zones under mid-ocean ridges. Let us assume for the moment that all of it that enters a melting zone melts and is removed from the mantle. We can ask, how much of the mantle comprises subducted oceanic crust at any given time? Initially, the fraction of the mantle, f, that is subducted crust will have been zero so, as there was no crust to remove, the rate of removal in melting zones would have been zero. Later, the rate of removal will be $f\phi$, where ϕ (Eq. (10.1)) is the rate at which material is passing through the melting zones. Early on, the rate of removal would have been small, so the amount of crust in the mantle would build up. As crust accumulated within the mantle, its rate of removal would increase. Eventually they might approach a balance, where the rate of removal equals the rate of addition.

This situation can be analysed in a similar way to the removal of primitive mantle, and it is explained in Section C.2. The result is that f will exponentially approach a maximum $f_m = \phi_c/\phi$. With a crustal density of 2900 kg/m³ and thickness of 7 km, this gives $f_m = 0.06$. The fraction at any given model time, t_m, is

$$f = f_m \left[1 - \exp\left(-t_m/\tau\right)\right]. \tag{10.7}$$

This specifies that f starts at zero, initially builds up at the rate $1/\tau_c$, but then the rate of accumulation slows and f approaches f_m. After 18 Gyr of model time, f will be 99% of f_m.

In other words, the mantle should contain about 6% of subducted, unprocessed oceanic crust. The reason the amount of crust within the mantle reaches a maximum is that old subducted crust is removed by melting within melting zones. Thus in the asymptotic state old subducted crust is removed as fast as it is added. As we will see later, not all crust melted under ridges may have been removed from the mantle, so the total amount of crust-derived material may be about double this.

This result quantifies the widespread expectation that, after several billion years of subducting oceanic crust plus underlying depleted mantle, the mantle will have substantial heterogeneity of major elements, with accompanying heterogeneity of trace elements.

10.5 Melting in a heterogeneous mantle

Probably the most important implication of major element heterogeneity is that the processes of mantle melting and melt extraction will be substantially different from those in the kind of uniform source that has been commonly assumed. Almost every aspect of these processes needs to be re-examined.

The main heterogeneity being injected into the mantle is the duality of oceanic crust and its complementary depleted mantle. The basaltic composition of oceanic crust forms eclogite at upper-mantle pressures, and eclogite melts at lower temperatures in the upper mantle than does a typical peridotite of average mantle composition, as is illustrated in Figure 10.11. Thus, as a heterogeneous mixture of eclogite and peridotite rises under a mid-ocean ridge, it is the eclogite that will melt first, and it may melt substantially before any peridotite melts. It may even produce more melt than a comparable homogeneous source [169].

Over the past decade petrological and geochemical studies have addressed the question of melting in such a heterogeneous source (e.g. [202, 204, 205] and references therein). A key factor is that melt derived from eclogite pods will be out of chemical equilibrium with the surrounding material, peridotite. The melt will

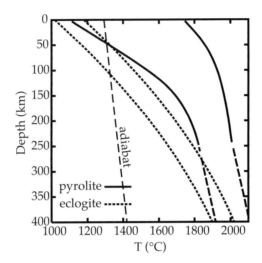

Figure 10.11. Solidi (lower) and liquidi (upper) for an average mantle composition (pyrolite) and for eclogite, the upper-mantle form of subducted ocean crust. The solidus is the temperature at which melting begins and the liquidus is the temperature at which melting is complete. An approximate solid adiabat is included. After Yasuda *et al.* [80]. Copyright American Geophysical Union.

therefore react with the peridotite to form an assemblage of intermediate composition, and it may solidify as it does so, yielding 'refertilised', hybrid lherzolites or pyroxenites [205–207]. If such material continues to rise under a mid-ocean ridge, it may remelt and mix with melt derived from melting peridotite. It is possible that the end-product of this process is a melt that is similar to the melt that would have been produced by a homogeneous source of the same average composition. However, we must consider that not all of the intermediate material may remelt, or that some of the eclogite-derived melt may reach the surface without fully mixing with peridotite melt.

On the one hand, the studies so far have yielded evidence that melt from heterogeneities does indeed contribute discernibly to MORBs and OIBs. On the other hand, they have produced arguments that melt from the heterogeneities often may not equilibrate with the surrounding peridotite. The same arguments suggest that significant amounts of melt may not be erupted at mid-ocean ridges, and thus may be trapped in the mantle and recirculate internally.

10.5.1 Melting a homogeneous source

Before discussing heterogeneous melting, it is helpful to depict the picture of melting under mid-ocean ridges that has developed from the assumption of melting of a homogeneous source, as sketched in Figure 10.12. As material rises

10.5 Melting in a heterogeneous mantle

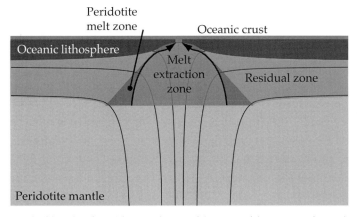

Figure 10.12. Sketch of melting under a mid-ocean ridge, assuming a homogeneous source. The curves are flow lines.

(passively) under the ridge, the reduction of pressure allows it to begin melting, or in other words its adiabatic temperature trajectory intersects the peridotite solidus (Figure 10.11, where pyrolite is similar to peridotite). This is estimated to occur at about 60 km depth.

The eruption zone at a mid-ocean ridge is very narrow, only about 10 km across. Yet melting will begin across a much broader zone. It is inferred that the rising melt is 'focused' into the narrow eruption zone, otherwise it is difficult to explain the thickness of the oceanic crust. The focusing occurs because the diverging flow under the ridge generates a pressure deficit that pulls the melt towards the ridge axis [208].

The rate of melting within the melt zone is proportional to the vertical velocity of the mantle material. As the off-axis flow lines curve towards the horizontal, the rate of melting will approach zero. A little of the off-axis melt may not be extracted to the ridge. Material that has passed through the melt zone will have had some melt extracted, forming a residue that will move laterally with the overlying plate. The degree of melting and depletion will be greater for material that reaches shallower depths. Off-axis melting will be limited at the top by the thermal boundary layer (the oceanic lithosphere) extending downwards by conduction (Chapters 5 and 6).

10.5.2 Reaction and disequilibrium of eclogite melts

Returning to melting in a heterogeneous source, the reaction of eclogite-derived melts with peridotite might produce a range of compositions, from lherzolite through pyroxenite, garnet pyroxenite even to eclogite, depending on the relative proportions of melt and peridotite. However, Sobolev *et al.* [209] propose that the product will be a relatively uniform pyroxenite, with approximately 50% inputs each from the eclogite melt and the peridotite. For the sake of conciseness, the term *hybrid pyroxenite* will be used here.

There is by now significant observational evidence that some melts from eclogitic or pyroxenitic sources reach the surface without fully equilibrating to the homogeneous composition. Takahashi *et al.* [210] concluded that the Columbia River Basalts are derived from shallow melting of a pyroxenite source in a mantle plume head. Sobolev *et al.* [209] argue that unusually high Ni and Si contents of Hawaiian shield basalts are consistent with derivation from a secondary, olivine-free pyroxenitic source produced when melt from recycled oceanic crust hybridises with peridotite in the Hawaiian plume [74]. Others have argued for some time that small near-ridge seamounts are produced by melting from heterogeneities, plausibly of recycled oceanic crust, that only pass through the edge of the sub-ridge melting zone [193, 194]. Salters and Dick [211] show that abyssal peridotites from the southwest Indian Ridge cannot explain the neodymium isotopes of nearby basalts without invoking a more enriched component, plausibly pyroxenite or eclogite, that has been completely melted out of the residual peridotites.

Osmium isotopes provide some of the strongest evidence for the survival in erupted basalts of unequilibrated signatures from eclogites [204, 209]. Osmium isotopes correlate nearly linearly with Sr, Nd and Pb radiogenic isotopes, and the sublinear correlations have been interpreted as indicating mixing between liquids, rather than reaction between a liquid and a solid. This would imply that the eclogite-derived melts survive their passage through the peridotite matrix until they reach the peridotite melting zone, or even near-surface magma chambers.

Kogiso *et al.* [204] have considered in some detail the physical circumstances in which eclogite-derived melts might survive both the initial melting process and then the passage through the peridotite matrix, taking account of the size of the eclogite body, diffusion rates in solids and liquids, and whether the melt is saturated or undersaturated in silica. Their general conclusion is that some eclogite-derived melt might reach the shallower peridotite melting zone if its source pods of eclogite are thicker than 1–10 m, whereas other melt will refreeze and be trapped in the mantle. Melts from smaller eclogite bodies or silica-undersaturated melts are the most likely to be trapped. The melt that does survive migration may do so by passing through relatively narrow channels that become insulated from surrounding peridotite by a reaction zone formed by preceding eclogite melt. A sketch of the resulting possibilities is shown in Figure 10.13.

If eclogite-derived melt passes through a channel without reacting with surrounding peridotite, then some of the chemical disequilibrium between the eclogite and the peridotite will remain. This means, on the one hand, that the eclogite-derived melt will retain some signature of its eclogitic source, which the observations just cited support. On the other hand, it means that some of the peridotite will not

10.5 Melting in a heterogeneous mantle

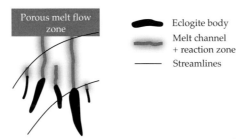

Figure 10.13. Sketch of the migration of melt emerging from eclogite bodies. The melt will react with surrounding peridotite, and will tend to refreeze. Some might react completely and become trapped in the mantle. However, melt from larger bodies might form a reaction zone that insulates later melt from reaction, and allows it to rise into the zone where pervasive peridotite melting will allow porous melt flow. After Davies [118]. Copyright American Geophysical Union.

reflect, in its trace element composition, the presence of eclogite in the source. This has important implications for the interpretation of mantle geochemistry.

10.5.3 Melt trapping and recirculation

An implication of Kogiso *et al.*'s study [204] is that some melt may react with the peridotite matrix, refreeze and not be remelted. Material close to the spreading axis is likely to ascend close to the Earth's surface and therefore to remelt and to be extracted. Any melt from heterogeneities that ascends into the zone where the peridotite matrix begins to melt is also likely to be extracted, as the peridotite melt will form a connected network through which the melt from heterogeneities can migrate.

However, eclogites and hybrid pyroxenites begin to melt much deeper than peridotite, about 110 km (Figure 10.11 and [80]), so their melting zone will be much wider. Heterogeneities further from the spreading axis may not rise shallow enough to remelt, nor may they rise into the peridotite melting zone. This material may be carried away laterally without remelting. A sketch of how melt may be trapped is shown in Figure 10.14. The eclogite melting that occurs outside the peridotite melting zone will tend to be in disconnected pockets. Melt from larger eclogite bodies may migrate, but much of it may not reach the surface.

Because the eclogite melting zone is deeper and wider than the peridotite melting zone, much of the eclogite melt, or its hybrid pyroxenite product, will pass outside the peridotite melting zone and may remain in the mantle. This effect will be enhanced if the peridotite matrix is more refractory than the 'fertile peridotite' usually assumed in models of the melting of a homogeneous source. This is because the refractory peridotite melting would begin shallower than 60 km and so be even less likely to capture the pyroxenite products.

176   Mantle chemical evolution

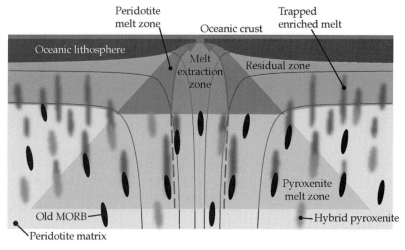

Figure 10.14. Sketch of melting under a mid-ocean ridge for a heterogeneous source. Old subducted oceanic crust (MORB) in the eclogite assemblage will start melting at a greater depth than peridotite. Some of the resulting melt may fail to be extracted, and remain as hybrid pyroxenite. Multiple generations of hybrid pyroxenite may return to the melting zone. After Davies [118]. Copyright American Geophysical Union.

An implication of this picture is that substantial amounts of hybrid pyroxenite will be recirculated within the mantle. Over time some of it will return to MORB melting zones, potentially to be remelted. Thus there will be multiple generations of hybrid pyroxenite, and a significant population of it will accumulate, as foreshadowed in Section 10.4.4. The material entering melting zones will have three main components, not two: peridotite residue, subducted oceanic crust and hybrid pyroxenite. Incompatible elements will be carried by both the subducted oceanic crust and the hybrid pyroxenite. The two types of heterogeneous inclusion are depicted in Figure 10.14.

10.6 Previous estimates of trace element abundance

The abundance of uranium and, by implication, other incompatible trace elements found in Section 10.2 is more than twice that of previous estimates. Several reasons can be found for why the previous estimates are too low, some relating directly to heterogeneity, others not.

10.6.1 Sampling heterogeneity, not mixing reservoirs

The Nd–Sr isotopic array in Figure 10.2(c) was initially interpreted as a mixing line between compositions of two reservoirs, a strongly depleted upper mantle and

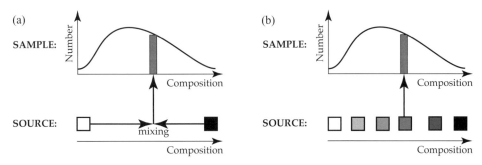

Figure 10.15. Contrasting conceptions of mantle sources. (a) Generation of a sample by mixing material from two reservoirs before eruption. (b) Direct sampling of a heterogeneity of intermediate composition.

a primitive lower mantle [107]. The concept is illustrated in Figure 10.15(a). In contrast, if the mantle is conceived of as being heterogeneous, with a range of compositions already existing in the mantle, then the Nd–Sr array can be viewed as directly sampling that range of compositions, as in Figure 10.15(b).

Some of the reasons for expecting the mantle to be heterogeneous have been covered in Section 10.4. In addition, we can note that Figure 10.2(a) would require at least three source types to span the two-dimensional array of data. Indeed, geochemists soon defined four or more source types required to span the data in higher dimensions [177, 178, 212]. Thus two-reservoir mixing is not plausible.

10.6.2 No primitive mantle

The identification of the less depleted source as primitive was always in conflict with the lead isotope data (Figure 10.3(a)), which implies that this reservoir would have ^{206}Pb/^{204}Pb of 22 or more, well away from the primitive 'geochron' locus, whereas the putatively depleted source plots around the geochron. This is the reverse of the relationship that would be expected, as I noted in 1984 [115]. There is no significant evidence for a primitive reservoir [60], and later trace element ratios such as Nb/U (Figure 10.7) reinforce the point made from the lead isotopes. Nor do we expect a significant amount of primitive mantle, according to the discussion of Section 10.4.3.

The later proposal for a deeper reservoir that is enriched but not necessarily primitive [108] avoids some of these problems. Indeed, its origin was ill-defined enough that it could be a receptacle for virtually any source type required to span the data. However, this layer would be required to contain half of the Earth's uranium and would be expected to generate a large plume flow, as discussed in Section 8.3 (see Figures 8.6 and 8.7), and the Earth's topography is inconsistent with this.

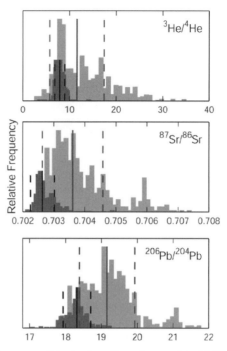

Figure 10.16. Histograms of isotopic ratios observed in MORB (dark grey) and OIB (light grey). Helium is normalised to the atmosphere. Dashed lines mark one standard deviation about the medians (solid lines). ^{3}He/^{4}He values for MORB and OIB are $7.92 \pm 1.06\,R_A$ and $11.62 \pm 5.82\,R_A$, respectively; ^{87}Sr/^{86}Sr values are 0.7026 ± 0.0004 and 0.7036 ± 0.0010; and ^{206}Pb/^{204}Pb values are 18.32 ± 0.39 and 19.16 ± 0.77. After Ito and Mahoney [213]. Copyright American Geophysical Union.

10.6.3 Focus on the depleted MORB mantle

The influence of the two-reservoir picture has persisted in a more subtle but quite misleading way, in the concept of the *depleted MORB mantle*. The idea that the MORB source has a relatively tightly defined composition that can be clearly separated from other putative components of the mantle has become so entrenched that it is difficult to find illustrations of the real distributions of data. One such illustration is shown in Figure 10.16. The populations of MORBs and OIBs completely overlap in all three systems. The MORB populations for Sr and Pb are rather skewed, with a fat tail extending to the enriched side of the distribution, though there are not enough data in this example to make this really clear.

Although, strictly, the depleted reservoir of a two-reservoir mantle ought to be more depleted than any observed value, some heterogeneity of the MORB source is conceded, and the depleted MORB source is taken to be represented by something called *normal MORB* (NMORB). NMORB is taken to be the most

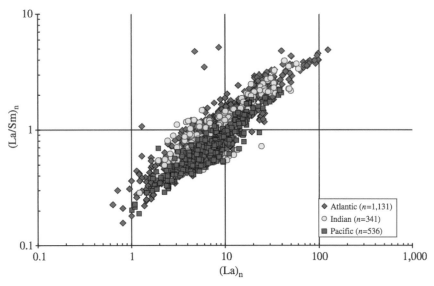

Figure 10.17. La/Sm, normalised to primitive mantle, versus La for MORB from three ocean basins. Lanthanum concentrations vary by about two orders of magnitude; La/Sm varies by more than one order of magnitude. Data were extracted from PetDB. From Hofmann [60]. Copyright Elsevier Science. Reprinted with permission.

common composition, or in other words the peak or mode of the distribution. For example, in Figure 10.16, NMORB strontium would be represented by the peak around 0.7025, presumed symmetric. Higher Sr ratios that do not fit this peak are then taken to represent other species, such as *enriched MORB* (EMORB).

However, there is no objective way to separate the NMORB and EMORB populations, as has been emphasised also by Hofmann [60]. Figure 10.17 shows La/Sm ratios for over 1000 MORB samples. This ratio is a measure of depletion or enrichment, since it effectively measures the slope of the trace element distributions in Figure 10.1. According to Hofmann, NMORB is often defined by La/Sm < 1, but in this plot such a division is completely arbitrary.

Furthermore, because high La/Sm correlates with a high La concentration, the exclusion of the high-La/Sm data will exclude much or most of the La content of the MORB source. This is true in other systems too. For example, the samples with high $^{87}Sr/^{86}Sr$ also tend to have higher Sr concentrations, so if the high-$^{87}Sr/^{86}Sr$ samples are excluded, much of the Sr inventory may be missed.

10.6.4 Mean compositions, not end-members

If the heterogeneity evident in the data of Figures 10.2, 10.3 and 10.16 is a direct reflection of the heterogeneity of the sources, as depicted in Figure 10.15(b), then

Table 10.2. *Average MORB composition estimates.*

	Nb (μg/g)	Th (ng/g)	U (ng/g)
'All MOR'[a]	5.95	400	240
NMORB[b]	2.33	120	47
EPR average[c]	3.79	200	80
EMORB[b]	8.3	600	180
MARK EMORB[c]	16.05	1100	305

[a] Raw averages from PetDB database, www.petdb.org/petdbWeb/.
[b] Sun and McDonough [188].
[c] Donnelly et al. [189].

the source regions should be characterised by a mean and spread, rather than by hypothetical end-members for which there is no good evidence.

This is most telling for the OIBs, particularly for helium, for which the two-reservoir model is still widely held. Thus there is held to be an 'undegassed' reservoir with high ^{3}He/^{4}He [214], which must be at least 35 R_A (atmospheric ratio) to accommodate the highest value in Figure 10.16. However, if the OIB distribution simply represents the values in a heterogeneous source, then the mean value is the most appropriate way to characterise the source. The mean value is only 11.6 R_A, less than twice the mean of MORBs, 7.9 R_A. The main distinction between the helium of MORBs and OIBs is not the difference in mean values but the difference in the spread, $\pm 1\,R_A$ for MORBs and $\pm 6\,R_A$ for OIBs.

Not only are the mean and spread a more accurate way to characterise the observations, but also the mean is the quantity required in mass balances and in considering mantle heat sources. Unfortunately, the means are not easy to estimate at present, partly because the mean may be strongly affected by relatively uncommon enriched samples, and partly because so little serious effort has been made to compile the required data. For example, the mean La/Sm value of Figure 10.17 should be calculated by weighting the individual values with their La content, which varies by two orders of magnitude. The mean will be strongly skewed to the high end.

A crude indication of the effect of including most MORB values is shown in Table 10.2, which lists abundances of Nb, Th and U. The first row shows simple raw averages from the category 'All mid-ocean ridges' from the PetDB database. The other rows show comparison values for two estimates of 'normal MORB' and two 'enriched' MORBs. In the all-MORB average U is 3–5 times larger than previous normal MORB estimates. It may be that U values are less reliable because U concentrations are so low, so Th and Nb values have been included. Typical

10.6 Previous estimates of trace element abundance

Table 10.3. *Plume contributions to mantle inventories.*
(a) Enrichments of plume components[a]

	Nb	Th	U
Hawaii/NMORB	2.2	2.9	2.4
EM1/NMORB	13	24	15
EM2/NMORB	11	42	24
HIMU/NMORB	17	28	18

(b) MORB-source enrichments due to plume components[b]

Plume enrichment	MORB-source enrichment
2	1.25
3	1.5
5	2
10	3.25

[a] Values taken from Figure 10.1. NMORB: 'normal MORB'; EM1: enriched mantle 1; EM2: enriched mantle 2; HIMU: high ^{238}U/^{204}Pb [177, 178].
[b] Enrichment factors are relative to 'depleted MORB mantle' inferred from 'normal MORB'. Plume material is assumed to comprise one-quarter of MORB-source material (see text).

MORB ratios are Th/U = 2.6 and Nb/U = 47 [60], so these are suitable proxies. All-MOR Th values are 2–3 times larger than previous estimates, while Nb values are 1.6–2.6 times greater. These factors may be minima, because plume-affected ridge segments may have been excluded from the 'all-MOR' category of PetDB, though this is not clear from the PetDB summaries.

10.6.5 Heterogeneities from plumes

A potentially more robust estimate of mean MORB-source composition, though with rather large uncertainties at present, comes from the composition of OIBs, inferred to be produced by plumes. In Section 10.4.1 it was estimated that about 25% of heterogeneities in the mantle may come from plume material that did not melt, and that has been stirred back into the mantle. This material could add significantly to the inventory of trace elements.

Table 10.3(a) shows the enrichments of various classes of OIBs relative to MORBs for Nb, Th and U. Hawaiian OIB is enriched by a factor of 2–3, whereas the other OIBs are enriched by factors of 10–40, with a mean of around 20. The higher enrichments are probably due in part to relatively low melt fractions [60], so they may overestimate the mean enrichments of the associated plumes.

Table 10.3(b) shows the effect of plume enrichments by factors of 2–10 on the mean composition of the MORB source. If plumes are enriched by factors of 3–5 relative to the 'depleted MORB mantle', then the mean mantle concentrations of incompatible elements is increased by factors of 1.5–2. Greater enrichments cannot be ruled out at this stage.

The accuracy of the latter estimates could be considerably improved by carefully estimating the melt fraction and enrichment of each plume and combining them with estimates of the melt volume (from erupted volumes) and the plume flow rate (from hotspot swells) [1], following the approach used in Section 10.4.1.

10.6.6 Chemical disequilibrium

Two recent estimates of the MORB-source composition [185, 186] use the composition of peridotites in key parts of their arguments. The trace element concentrations estimated by Salters and Stracke [185] are deduced from major elements through a correlation between lutetium, a compatible trace element, and CaO in anhydrous spinel peridotites. However, we would expect the MORB source to have contained eclogitic or pyroxenitic heterogeneities, and they would be expected to be enriched in trace elements. By the arguments in Section 10.5.2, the surrounding peridotite may not have equilibrated with the higher Lu content of the heterogeneities. Thus the Lu in the peridotites may well underestimate the Lu content of the whole source. All of the other trace element concentrations are tied to the Lu content, so the trace element content of the source would be underestimated.

The estimate by Workman and Hart [186] starts from the trace element composition of clinopyroxenes from abyssal peridotites. Those peridotites likewise may not have equilibrated with enriched heterogeneities, so their estimates may also underestimate the trace element content of the whole source.

It is notable that Salters and Dick [211] found that isotopes in some MORBs from the Indian Ocean could not be explained as products of nearby abyssal peridotites, but implied an additional enriched component. They note that pyroxenites and eclogites are rare at mid-ocean ridges, though common among xenoliths and peridotite massifs, and inferred that enriched heterogeneities had been melted out of the source before it reached the surface. In that case the residual peridotites would fail to represent the incompatible elements in those heterogeneities.

These estimates are also vulnerable in other respects. They both rely on equilibrium melting models for a homogeneous source, and both use long chains of inference. For example, Salters and Stracke's [185] estimate of thorium content depends on a chain of no less than eight elemental ratios. U and K require a further ratio each. Workman and Hart [186] begin with a reconstruction of whole-rock composition, which requires partition coefficients and modal abundances, then a melting model to estimate pre-melt composition. Both studies focus on establishing the composition of what they call the depleted mantle, to the exclusion of more enriched components.

Thus it may be that the 'depleted MORB mantle' is more enriched than either of these estimates. This would imply that both enriched heterogeneities and the mean composition of the MORB source would then be proportionately more enriched.

10.6.7 Basaltic gradient through the upper mantle?

A final reason why previous estimates of the trace element content of the mantle may be too low comes not from geochemistry but from the numerical models of mantle stirring discussed in Chapter 9. Figure 9.8 reveals a gradient in the concentration of basaltic tracers through the upper mantle. The tracers, representing subducted oceanic crust, which is denser than average in the upper mantle, tend to settle through the upper mantle, leaving the top of the upper mantle depleted by about a factor of 2 relative to the lower mantle. The effect is strongest in the early, hot mantle, but it persists into the present in a number of models like those in Figures 9.5 and 9.10. This mechanical depletion reduces the amount of melting and results in the oceanic crust being much thinner than if the upper mantle had the average basaltic content.

The persistence of this effect into the present was a little surprising, because an earlier model run at present conditions (in other words, with steady heating rather than heating that declines with radioactive decay) showed no gradient in the upper mantle [123]. Therefore the behaviour of several evolving models was checked, with results shown in Figure 10.18. The evolving models all show a clear gradient through the upper mantle, whereas the steady model shows very little gradient. The steady model also did not develop an accumulation at the base of the mantle.

The final degree of depletion is not easy to discern from Figure 10.18. It is better represented by the comparisons of actual crustal thickness and 'fertile' crustal thickness shown earlier (Figures 9.8 and 9.11). These show that the final crustal thickness is 70–90% of the fertile thickness, suggesting a depletion of 10–30%.

The difference in basal accumulation among the models is easier to understand. Evidently the lower mantle is viscous enough under present conditions to prevent significant settling to the base of the mantle. On the other hand, settling can occur in the early, hot mantle when the lower mantle is less viscous. Once an accumulation

184   *Mantle chemical evolution*

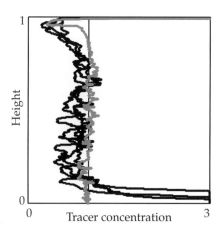

Figure 10.18. Profiles of concentration of basaltic tracers. The black profiles are from evolving models at the end of the run, including the models of Figures 9.5 and 9.10 and Case 1 from Davies [156]. The grey profile is from the steady model 'Coolr' from Davies [123].

is established, it will heat up and its viscosity will be much lower than the overlying mantle, which would make entrainment of the accumulation harder. Therefore the accumulation persists into the present, even though it could not form now. It is not clear why the upper-mantle gradient should persist into the present, because subduction should keep the upper mantle well connected and stirred with the lower mantle.

Thus the evolving models indicate that some mechanical depletion of the uppermost mantle may occur. This would mean that the concentrations inferred from melting under mid-ocean ridges were lower than in the deeper mantle. The effect could require the mean mantle trace element composition to be revised upwards by as much as a factor of 2, though there is still some uncertainty about the magnitude of the effect.

10.7 Dynamical modelling of refractory incompatible elements

The observations summarised in Section 10.1 establish four things about the refractory incompatible elements in the mantle. First, there is a significant degree of heterogeneity in both trace elements and isotopes. Second, OIBs are distinctly more enriched and more heterogeneous than MORBs (Section 10.3). Third, both populations reveal an apparent age of about 1.8 Ga in the lead isotope plot (Figure 10.3(a)). Fourth, there is no clear evidence for a primitive component.

The mantle abundances of heat source and other trace elements inferred in Section 10.2 are consistent both with global budgets of refractory elements and with the geophysical requirement that the mantle is heated more within than from below (Chapters 7 and 8). The significantly lower concentrations obtained in previous estimates for the 'depleted MORB mantle' are plausibly explained by the failure to consider the full implications of mantle heterogeneity and by a continuing tendency to think in terms of mixing from end-member reservoirs or components, as discussed in Sections 10.4 to 10.6. The remaining significant discrepancy is between the heating required by conventional thermal evolution models (Sections 9.1 and 9.2) and the global abundances of heat source elements inferred from cosmochemistry, as discussed in Section 9.3.

Dynamical models of the mantle, of the kind presented in Chapter 9, have by now reproduced the four main features of the refractory incompatible elements. We can therefore consider that the physical and chemical models of the mantle have been reconciled regarding the important features of these elements, except for the level of radioactive heating required for evolving models. The basis for this conclusion will now be outlined.

10.7.1 Heterogeneity and the MORB–OIB difference

Christensen and Hofmann [122] showed the way in 1994 by modelling subducted oceanic crust and depleted mantle in a convection model using tracers. They showed that some crustal material accumulated at the bottom, thus supporting Hofmann and White's hypothesis [74] that plumes incorporate additional oceanic crust from the D'' layer. They showed that residence times would yield apparent ages similar to the lead observations. They also obtained a range of isotopic heterogeneity comparable to the observations. Their model was limited in significant respects, both by the computer power available at the time and by some of their assumptions, but they established an approach that has proven very fruitful. Subsequent modelling extended this approach to the full mantle Rayleigh number and full age of the Earth [203], to three dimensions [215–217] and to fully evolutionary models [156, 218] like those of Figures 9.5 and 9.10.

Davies [203] discussed the approach and added several considerations. The Christensen and Hofmann [122] (C&H) models were run for only 3.6 Gyr at steady conditions, but about 18 Gyr are required at steady conditions to achieve the required number of transits (Section 10.4.3), or evolving models with decaying heat sources are needed, as later became feasible. Most of the C&H models were mainly bottom-heated, rather than mainly internally heated, and the resultant strongly upwelling sheets may have affected residence times. C&H sampled the chemical tracers throughout the model at the end of the run, whereas geochemists sample only the oceanic crust, formed at the top, and OIBs, drawn from the bottom, so the models should be sampled accordingly.

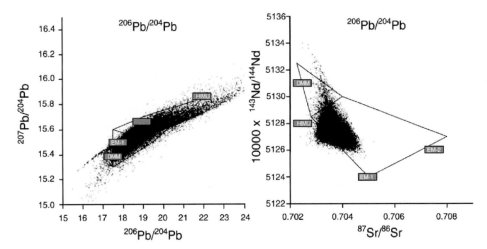

Figure 10.19. Example of synthetic isotope plots from a numerical model of mantle dynamics with tracers. The lines and boxes summarise the observations. The spread of isotope values is sensitive to poorly constrained sampling scales and melting processes, so these results should only be taken as indicative. From Brandenburg et al. [219]. Copyright Elsevier Science. Reprinted with permission.

Finally, the synthetic isotope plots constructed from the final model results are very sensitive to several poorly constrained factors, including the sampling volumes in the mantle and the degree to which melts mix, which will have a large effect on the isotopic spread. To those we can add the effects of heterogeneity and associated disequilibrium, whereas the estimates assume homogeneity and equilibrium, and even then a number of the required geochemical parameters are not well determined. Thus, although it is possible to obtain plausible synthetic isotope plots from the models, they should not be regarded as strong tests of the mantle models. They are, to a greater degree, tests of our understanding of melting and melt migration. A more recent example of synthetic isotope plots is shown in Figure 10.19.

C&H found that the apparent age generated in their synthetic lead isotope plot (1.6–1.8 Ga) was significantly larger than the residence time of tracers in the model (about 1.4 Ga). The reason is that the apparent age is strongly controlled by the oldest tracers, which generate the broadest spread of values and thus weigh more heavily in constraining the linear correlation. Thus models that generate residence times of 1.4–1.6 Ga can be considered consistent with observations for the purpose of the following discussion.

Because of these considerations, the most robust observational features are the differences between MORBs and OIBs and the apparent lead age, corrected downwards for comparison with residence times in models. The spread in isotopic ratios is a less reliable measure of mantle sources.

10.7 *Dynamic models of refractory incompatible elements* 187

Models that include tracers of dense subducted oceanic crust usually form accumulations at the base, as in Figures 9.5 and 9.10. The tracer concentration profiles of Figure 10.18 also reveal the accumulations. These results, and those of others [218, 219], confirm the results of Christensen and Hofmann [122] and support the hypothesis of Hofmann and White [74] that plumes acquired an additional complement of basalt-composition material from the D'' layer. These models therefore offer a straightforward explanation for the enrichment of OIBs relative to MORBs.

Although this explanation is plausible, it is not easy to quantify, in terms of how much additional basaltic component a plume is likely to carry. There will be a physical limit, because the negative buoyancy of the basaltic component will counter the positive thermal buoyancy of the plume, and if there is too much basaltic component the plume would no longer ascend. There have been some instructive models of thermochemical plumes ([81–84]; Chapter 7), but their behaviour is not simple, as there is a tendency for heavy and light components to separate. Nevertheless, it ought to be possible to establish what amounts are plausible, and whether they are consistent with the observed enrichments of OIBs.

10.7.2 Residence times

The extraction of reliable ages from such models has required a couple of issues to be clarified. Christensen and Hofmann [122] obtained residence times of about 1.3 Gyr, but the models were run for only 3.6 Gyr at steady conditions, leaving the possibility that longer runs would have yielded larger residence times. Davies [203] ran models for 18 Gyr at steady conditions and obtained residence times up to 2.7 Gyr. On the one hand, this seemed to dispose of the previously perceived difficulty of mantle heterogeneities surviving for billions of years. On the other, it raised questions as to the source of the large residence times, which might have been due to using better plate simulations, to having a high-viscosity lower mantle (which the C&H models lacked) or to different assumptions about melting.

Subsequent studies [215–217] have demonstrated that residence times and degree of processing are controlled almost entirely by the rate at which material is processed through mid-ocean ridge melting zones (ϕ in Eq. (10.1)) or, equivalently, by the processing time τ (Eq. (10.2)). Other factors, such as the stiffness of subducted plates, the viscosity of the lower mantle, the vigour of convection (in other words the Rayleigh number), the dimensionality (two-dimensional versus three-dimensional) and the presence of toroidal flow (horizontal shearing) have only secondary effects. An example of a three-dimensional model with heavy tracers is shown in Figure 10.20.

It is important, however, to run the models for a sufficient time. Figure 10.21(a) shows that the average residence times of tracers are still increasing even after

188 Mantle chemical evolution

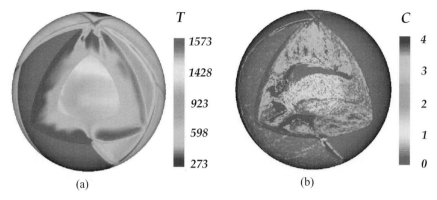

Figure 10.20. Grey scale versions of cut-away images of (a) temperature and (b) tracer concentration in a three-dimensional spherical model of mantle stirring with heavy tracers. After Huang and Davies [216]. Copyright American Geophysical Union.

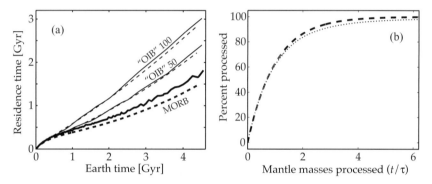

Figure 10.21. (a) Residence times from MORB and 'OIB' samples of three-dimensional models (solid curves). Dashed curves are estimates based on the sampling theory (Sections 10.7.3 and 10.7.4). 'OIB' 50 is for an excess density of 50 kg/m^3, and 'OIB' 100 is for 100 kg/m^3. (The 'OIB' interpretation is discussed in Section 10.7.4 below.) (b) Percentage of mantle processed through melting zones versus time, normalised by the processing time, from a 3D numerical model (dotted) and the sampling theory (dashed; Section 10.4.3). After Huang and Davies [215, 216]. Copyright American Geophysical Union.

the equivalent of 18 Gyr of processing at present rates. The amount of mantle processed, the complement of the amount of primitive mantle remaining (Eq. (10.5)), must also be run for several processing times to approach a realistic value (Figure 10.21(b)). Because convection runs nearly 10 times faster early in Earth history (Figure 10.10), models that are run at steady rates for only 4.5 Gyr or less will have accounted for less than the first billion years of evolution, and will seriously underestimate the degree of processing and the residence times.

10.7 Dynamic models of refractory incompatible elements

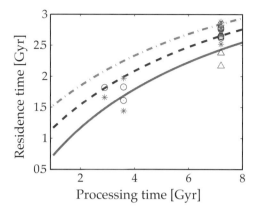

Figure 10.22. Final residence time as a function of processing time, τ. Curves are from the sampling theory (Section 10.7.3) for simple sampling (solid) and two cases in which a sampling delay has been assumed (dashed). Results from some numerical models are indicated by the symbols. From Huang and Davies [215]. Copyright American Geophysical Union.

A second requirement is to determine an accurate residence time for the mantle. This depends on the average plate velocity and on the depth of melting (Eqs (10.1) and (10.2)). The biggest uncertainty is the depth of melting. Because of mantle heterogeneity, trace elements are likely to be scavenged from a depth of about 110 km under mid-ocean ridges (Figure 10.11), and we saw in Section 10.4.3 that this gives a processing time of $\tau = 4$ Gyr. If the melting depth were taken to be the conventional peridotite melting depth of 60 km, then $\tau = 7.3$ Gyr and processing would be relatively slow. The relationship between residence time (at the end of a run) and processing time is shown in Figure 10.22, as estimated by the sampling theory of the next section. This indicates that a processing time in the range 3–4 Gyr will give residence times in the range 1.5–2 Gyr. However, there are some details that affect this result, as will be discussed in the next section.

These plots explain the main differences among earlier results. For example, Davies [203] in two models used an areal spreading rate of only 2.4 km²/yr (6 cm/yr linear spreading rate and an assumed ridge length of 40 000 km) and a melting depth that decreased to 70 km at present, and this gives a processing time of 7.2 Gyr. The mean MORB residence times from these models were 2.7 Gyr and 2.6 Gyr, which is within the range predicted in Figure 10.22. On the other hand, Christensen and Hofmann's [122] processing time was only 1.5 Gyr and their mean residence time less than 1.4 Gyr, also within the range predicted in Figure 10.22. Considering the noble gases in the mantle, van Keken and Ballentine [220, 221] estimated that the mantle would have been processed and degassed only by 35–70%, but their models were run at present rates for only 4 Gyr and for only 0.2–1.3 processing

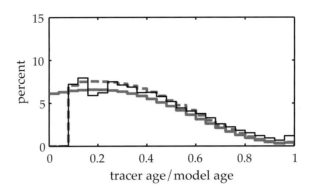

Figure 10.23. Histogram of MORB tracer ages (thin black) compared with the age distribution predicted from the simple sampling theory (solid grey) and with a sampling delay time of 0.4 Gyr (Earth time) (dashed grey). The tracer ages are given relative to the final model time. After Huang and Davies [215]. Copyright American Geophysical Union.

times. From Figure 10.21(b) we would expect 20–70% of the mantle to have been processed, which is consistent with their results.

10.7.3 Sampling theory

The simple sampling theory that predicts an exponential decline of remaining primitive mantle (Section 10.4.3) and an asymptotic proportion of subducted oceanic crust (Section 10.4.4) was extended by Huang to estimate the mean residence times of tracers and even their age distributions [215–217]. The mean residence time as a function of model time, t_m (i.e. equivalent to a model running at present rates, see Figure 10.10) is

$$t_R = \tau[1 - \exp(t_m/\tau)]. \tag{10.8}$$

This gives the solid curve of Figure 10.22 (for $t_m = 18$ Gyr, i.e. after 4.5 Gyr of evolution in Earth time).

Usually ridges are well separated from trenches, and it takes some time before any subducted material reaches a ridge melting zone. This will cause a delay before a tracer is processed, and a corresponding lack of young tracers. This was observed in the age distributions of tracers [203], an example of which is shown in Figure 10.23. This effect increases the processing time in the model. The theory was extended to take account of this by adding a delay time to the processing, and the grey curves in Figure 10.23 show predicted age distributions without (solid) and with (dashed) a processing delay, in this case of 0.4 Gyr. Mean ages with processing delays are included in Figure 10.22 (upper curves) for delays of 0.4 Gyr and 1.2 Gyr [215]. The numerical model results are mostly bracketed by these curves.

Huang [217] developed a further extension of the sampling theory to account for numerical models in which the melting depth decreases with time, according to the decrease in mantle temperature [203, 217]. The processing rate then decreases with model time, and decreases faster with Earth time. Results show that the degree of processing (as in Figure 10.21(b)) depends mainly on the early processing rate, whereas the final tracer residence times (as in Figure 10.21(a)) depend mainly on the late processing rate.

10.7.4 OIB ages

The 'OIB' ages shown in Figure 10.21(a) are significantly larger than the observed OIB apparent ages of about 1.8 Gyr. They are in quotes because the sample is, in retrospect, not really a good simulation of OIB sampling. In the numerical models the 'OIB' samples were simply taken from a thin zone at the bottom of the model [203, 215–217], assuming plumes would sample that region. However, when there are denser accumulations of subducted oceanic crust, plumes will come from the top or sides of the accumulations. There is considerable structure within the accumulations (Figures 9.5 and 9.10), with a thinner dense layer overlain by a more gradational zone generally with much lower concentrations of tracers. It is plausible that plumes would sample those outer zones rather than the denser zone at the bottom, and it is also plausible that the dense bottom zone would have larger residence times.

An indication of this was obtained by Davies [156] from several evolving models including those of Figures 9.5 and 9.10 (Cases 1 and 2 below). Table 10.4 shows horizontally averaged ages (residence times) from various depth ranges near the bottom of the model, compared with the nominal MORB and 'OIB' ages. The MORB age is from those tracers present in the oceanic crust at the end of the run, and the 'OIB' age is from the lowest 20 km of the model. Table 10.4 shows that the mean ages are greatest in the lowest 200 km, and decrease at higher levels. The ages in the 'piles' extending up to 1000 km are still significantly greater than the MORB ages.

Thus, for these models to be consistent with observed OIB ages, which do not seem to be notably larger than MORB ages, plumes would have to incorporate only a modest fraction of the basal tracers. We should bear in mind, however, that Chase [222], who was the first to highlight the age significance of the lead isotope plot (Figure 10.3) in 1981, separated the various OIBs and found apparent ages ranging from 1 Ga to 2.7 Ga. This suggestion ought to be revisited with more modern data.

Huang [216] further extended the sampling theory to address this in a different way, by assuming a population of tracers in the bottom accumulation that never leave once they have arrived. The proportion of such tracers was estimated

Table 10.4. *Ages (in Ga) from numerical models in specified depth ranges [156].*

Case	1	2	3	4
Mean age, MORB (model crust)	1.63	1.47	1.61	1.50
Mean age, 'OIB' (0–20 km)	2.97	2.54	2.59	2.45
Mean basaltic ages of deep layers:				
0–50 km	3.01	2.54	2.59	2.48
50–100 km	3.03	2.54	2.58	2.53
100–200 km	2.99	2.36	2.55	2.52
200–500 km	2.68	2.21	2.25	2.38
500–1000 km	2.27	2.13	2.23	2.09

from the rate at which the tracer concentration increases in the bottom zone, and their age distributions and mean ages were calculated, the latter being included in Figure 10.21(a).

To account for the lower observed ages of OIBs compared with model 'OIB' ages, Huang assumed that entrainment into a plume is inversely proportional to tracer concentration, as a proxy for density and following the theory of Sleep [223]. He estimated that in the conditions of the model a plume would entrain about 14% of tracers from the accumulation, so the OIB age would be 1.98 Ga, relative to model MORB and 'OIB' ages of 1.90 and 3.02, respectively. This value is sufficiently close to the model MORB age that the difference would not be very noticeable.

These results indicate that the models are reasonably consistent with observed OIB ages, though more accurate tests are desirable.

10.7.5 Heterogeneities remelted, not homogenised

There is a considerable literature devoted to modelling the rate at which chemical heterogeneities are stirred by convective flows (e.g. [199, 201]) as briefly discussed in Section 10.4.2. The motivation for this work was the idea that the apparent lead age reflects the time it takes for heterogeneities to be stirred down to a thickness at which it is effectively homogenised. This could be either the scale at which solid-state diffusion chemically equilibrates the material, which may be only centimetre scales [197], or, more likely, a scale at which melting and melt extraction will homogenise the chemical signals, which could be kilometres or larger. The challenge was to show how heterogeneities could survive mantle mixing for billions of years, which at first seemed intuitively implausible to many, as discussed earlier.

However, in the class of models pioneered by Christensen and Hofmann [122], the heterogeneities are removed by remelting, rather than by being stirred down to a scale at which the material or melt is homogenised. The empirical success of such models that has been summarised here indicates that this is the main way in which heterogeneity is removed in the mantle. The apparent lead age is then to be interpreted as a mean residence time, not a homogenisation time. The residence time is the time interval between successive passages through a mid-ocean ridge melting zone.

Stirring and homogenisation are strongly sensitive to such factors as the time dependence or dimensionality of the flow [198, 224], and this has motivated much of the modelling of this process. However, removal by remelting is evidently much less sensitive to such factors because it depends just on material being carried into or settling into the two sampling zones, and not on the detailed geometry or topology of the heterogeneities. This conclusion is supported by the consistency of results from models with various Rayleigh numbers, various viscosity structures, two-dimensional or three-dimensional, flat or spherical, and the presence or absence of plates, all of which would strongly affect the homogenisation process.

This implies that the results obtained here are fairly robust relative to the various approximations in the models, and this improves our confidence in them.

10.8 Resolving the noble gas enigma

The noble gas observations have been particularly difficult to interpret, for two main reasons. First, some, but not all, OIBs have unradiogenic helium isotopic compositions (meaning high ^{3}He/^{4}He or low ^{4}He/^{3}He; Figures 10.4 and 10.16) compared with MORBs. This has been taken to mean that the source is enriched in 'primitive' helium – in other words, that it has a higher abundance of the non-radiogenic ^{3}He. However, lead isotopes in particular (Figure 10.3) make it clear that these samples have been previously processed in melting episodes. The helium would be expected to degas during melting, and the 'primitive' helium to be lost. Second, estimates of the global budget of ^{40}Ar, which is the daughter of radioactive ^{40}K, indicate that about half of it is in the atmosphere, not much is in the continental crust, and so the balance, about half, should be in the mantle [225]. However, the MORB source has usually been inferred to have been strongly degassed and to have a low abundance, insufficient to balance the global budget.

The unradiogenic OIB helium has been a problem for all models, geophysical or geochemical. The argon mass balance has been dealt with by geochemists by assuming a large 'undegassed' reservoir somewhere deep in the mantle, initially identified with the allegedly primitive lower mantle [226]. This would accommodate all the argon (and other elements) required by awkward mass balances, but it is not compatible with the physical and dynamical interpretations of the mantle, as we have seen. The undegassed reservoir has also been claimed to explain the presence of unradiogenic helium in OIBs [226], but it requires plumes to acquire a small complement of unradiogenic helium from the reservoir without also acquiring an unradiogenic lead signature. Given also the requirement that plumes acquire a substantial extra complement of basaltic composition from somewhere, these are delicate operations that are not easy to quantify. Anyway, the idea of a large separate reservoir in the deep mantle is not dynamically tenable, so we need to move on from it. All we have available is the D'' layer and its associated large thermochemical piles.

The consideration of a heterogeneous mantle, and of melting heterogeneous sources, has led to an alternative interpretation that readily accounts for the helium enigma and may also account for the argon mass balance when all relevant uncertainties are taken into account. This will now be described, and some alternative interpretations will then be briefly discussed.

10.8.1 Noble gases in the hybrid pyroxenite

Section 10.5.2 reported investigations of melting in heterogeneous sources that indicate the formation of *hybrid pyroxenite*, produced when melts derived from eclogitic bodies react with surrounding, more abundant, peridotite. In Section 10.5.3 it was argued that significant amounts of such hybrid pyroxenite would not be extracted at mid-ocean ridges, and would therefore recirculate within the mantle. Successive generations of hybrid pyroxenite would form, as one generation is carried back into a melt zone, to contribute to the formation of another generation.

Incompatible elements, including noble gases, would be concentrated into hybrid pyroxenites, as they are into all melts. If some of the hybrid pyroxenite recirculates within the mantle, then it will not be degassed, so this is a way in which gas-rich bodies could form in the mantle. If such processes have operated since early in Earth history, before much of the noble gases had been lost from the mantle, then the hybrid pyroxenites would have acquired significant noble gases. As old subducted oceanic crust and hybrid pyroxenite are carried into a melting zone, both will melt preferentially compared with the peridotite in which they are embedded. Some of the melts will refreeze, as depicted in Figures 10.13 and 10.14.

Old oceanic crust and hybrid pyroxenite are likely to be intermingled after some mantle stirring, so some mixing of the two kinds of melt is to be expected as the melts migrate in the melting zone (Figure 10.24(a)). As already argued, some of the total melt will be erupted to form oceanic crust, and this will degas. On the other hand, some of the total melt will not be extracted, but rather will recirculate

10.8 Resolving the noble gas enigma

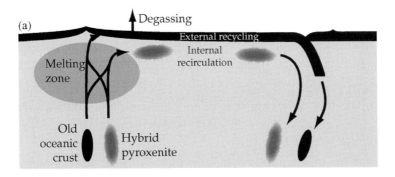

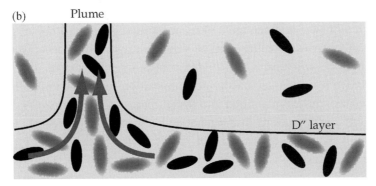

Figure 10.24. (a) Sketch of how hybrid pyroxenites may partially recirculate internally, while some is erupted to contribute to the externally recycling oceanic crust. Noble gases, carried mainly in the pyroxenites, would thus be only partially degassed at mid-ocean ridges. (b) Sketch of how both old oceanic crust and hybrid pyroxenite may accumulate in greater concentrations in the D″ zone. From Davies [227]. Copyright American Geophysical Union.

within the mantle, and so will not lose its noble gases (Figure 10.24(a)). Some of the gases initially present in the hybrid pyroxenite will thus be lost, but not all. The next generation of hybrid pyroxenite would still be the main site in which noble gases are stored in the mantle, even though it would have a lower concentration of noble gases.

So far this just describes one way in which the noble gases initially in the mantle are stored and gradually lost over the age of the Earth, although it does mean that previous estimates of their present concentrations may be inaccurate, as we will see. The real puzzle has been why some OIBs, coming from sources that show obvious signs of having been previously processed, have less radiogenic noble gases than MORBs.

Pyroxenites, being closer in composition to eclogites than are peridotites, have similar mineral phases as eclogites, though in different proportions. They are therefore likely to be denser than average mantle through most of the mantle depth, as is eclogite and its deeper equivalents. This means that they will behave dynamically in a similar way to the subducted oceanic crust modelled, for example, in Figures 9.5 and 9.10. Therefore hybrid pyroxenites will tend to accumulate in the D'' region, along with subducted oceanic crust, and both will be prone to entrainment by plumes, as depicted in Figure 10.24(b). Thus plumes would deliver material (old oceanic crust) whose refractory elements and isotopes would record previous processing, as well as material (hybrid pyroxenite) that has a significantly higher complement of noble gases.

The final piece of the puzzle is that the material in D'' has a longer residence time than that in the mantle interior, as is evident in Table 10.4 (Section 10.7.4). This means that pyroxenite that spends time in the D'' accumulation will degas less often, and it will have retained more of its primordial gases. The rate of generation of radiogenic isotopes, like ^{4}He, will be similar in pyroxenites wherever they are, but those in the D'' will tend to have higher non-radiogenic isotopes, like ^{3}He. Thus they will have a less radiogenic isotopic composition, such as a lower ^{4}He/^{3}He.

This scenario can be quantified and tested, but to do so we need estimates of the noble gas concentrations in the mantle. These need to be reconsidered for a heterogeneous mantle, but first we will briefly review conventional estimates. A more detailed discussion is given by Davies [227].

10.8.2 Conventional concentration estimates

Abundances in the MORB source have previously been based on the flow of ^{3}He from mid-ocean ridges. The degassing rate of ^{3}He was estimated by Farley *et al.* [228] to be 1060 mol/yr, based on observed concentrations of ^{3}He in the ocean combined with ocean circulation models. Although this value is given to three significant figures, we should bear in mind that there are likely to be significant uncertainties, both from the ocean circulation model used and from potential time variations in such a remarkably small global quantity.

Assuming helium is efficiently extracted by the magma that forms the oceanic crust, its concentration in the source follows from the degassing rate [225, 229]. Thus about 20 km^3 of oceanic crust is formed each year. If for the moment we follow the usual estimate that MORB is produced by 6–10% melting in the MORB source [186, 190], then the annual volume of MORB (3 km^2/yr times 7 km thickness) comes from a volume of about 200–333 km^3. This would imply a concentration of $(6–10) \times 10^8$ atoms ^{3}He/g. The observed ^{4}He/^{3}He ratio of 90 000 then implies a ^{4}He concentration of $(5–9) \times 10^{13}$ atoms ^{4}He/g. These values are summarised in Table 10.5(a) as the 'Low gas' case.

Argon abundances have been obtained through the ^{4}He/^{40}Ar ratio. Allegre *et al.* [225] quote an observational range of 2–15. Porcelli and Wasserburg [181] assume

Table 10.5. *Noble gas abundances. Abundances in atoms per gram. (Plausible ranges in parentheses.)*

	^{4}He (10^{13})	^{3}He (10^8)	^{21}Ne (10^6)	^{22}Ne (10^8)	^{40}Ar (10^{13})	^{36}Ar (10^8)
(a) Conventional estimates, MORB source						
Low gas	7	8	6	0.9	2	7
	(5–9)	(6–10)	(4–8)	(0.6–1.1)	(0.5–3)	(2–10)
(b) New estimates, MORB source						
Medium gas	14	16	12	1.8	15	~30?
High gas	20	24	18	3	22	~70?
(c) 'Undegassed lower mantle'						
Porcelli and Wasserburg [181]	100	500	570	150	56	600

that these isotopes reflect the amount generated over a residence time of about 1.4 Gyr in their steady-state upper mantle, and use a production ratio of about 3.6, based on the usual K/U ratio of 1.3×10^4 [184]. The residence times given in Section 10.7.2 are comparable, so we can use their production ratio. Thus we get ^{40}Ar $= 2 \times 10^{13}$ and, with the usual ^{40}Ar/^{36}Ar ratio of about 30 000, ^{36}Ar $= 7 \times 10^8$ (Table 10.5(a)). Ballentine *et al.* [229] use a production ratio of 1.55, appropriate for 4.5 Gyr, which probably overestimates the ^{40}Ar abundance.

Neon follows by similar arguments, from a production ratio for ^{21}Ne/^{4}He of 0.45×10^{-7} [230] and the ^{21}Ne/^{22}Ne MORB ratio of 0.07 (taking account also of the initial solar ratio of 0.0328).

Abundances in the lower mantle of the old two-layer mantle were built around the assumption that it retains all of the radiogenic isotopes generated within it over 4.5 Gyr. Values from Porcelli and Wasserburg [181] are reproduced in Table 10.5(c) as representative of this approach.

10.8.3 New concentration estimates

The abundances of noble gases in the MORB source might not be as well determined as the above estimates imply. Ballentine *et al.* [229] question whether the ~1000 year timescale implicit in ocean sampling of the helium flux from mid-ocean ridges is long enough to yield a reliable average, given possible fluctuations in plate tectonic processes on longer timescales. They cite two other estimates that give higher concentrations of helium. The gas-rich 'popping rock' [231, 232] has a ^{3}He concentration of 2.7×10^{10} atoms/g. With an assumed 10% partial melting, this implies a minimum concentration in the MORB source of 27×10^8 atoms/g.

The observed $CO_2/^3He$ ratio implies a source concentration in the range $(11–45) \times 10^8$ atoms/g.

Our consideration of melting a heterogeneous source also suggests higher concentrations than the conventional estimates. Because eclogite (and pyroxenite) is more fusible than peridotite (Figure 10.11), MORBs may comprise melt derived substantially from eclogites and pyroxenites. In Section 10.4.4 it was estimated that the mantle ought to comprise about 6% subducted oceanic crust, based on current subduction rates, and the arguments of Section 10.5.3 suggest that a comparable amount of pyroxenite would also be present. Thus 100% melting of the eclogite and pyroxenite could produce around 12% bulk melt fraction without any input from peridotite. This suggests that the peridotite is more refractory than the conventional homogeneous peridotite. In that case, rather than melting at 60 km depth (Figure 10.11), it would melt at a shallower depth. Because melt extraction will be inefficient outside the peridotite melt zone, this would increase the amount of melt that recirculates as pyroxenite. Thus the zone from which melt is mainly extracted would be smaller than that usually assumed, which would mean the concentration of noble gases would have to be larger to yield the observed helium flow.

If the refractory peridotite did not begin melting until 30 km depth, then the degassed helium would come mainly from above this depth and the implied concentration of helium in the MORB source would be double the usual estimate. If the peridotite did not melt until 20 km, the implied concentration would be tripled. The resulting ^3He concentrations of $(16–24) \times 10^8$ atoms/g are listed in Table 10.5(b) as the 'Medium gas' and 'High gas' cases. The corresponding values for neon are obtained through the same logic. These are more consistent with the estimates made by Ballentine et al. [229].

In the course of arguing for what they call a 'zero-paradox' mantle, Ballentine et al. [229] also argue that the MORB source should contain the ^{40}Ar that cannot be accounted for by the atmosphere (99×10^{40} atoms). If Earth has a K content of 240 μg/g [140], then it should have produced 194×10^{40} atoms of ^{40}Ar. The balance, 95×10^{40} atoms, gives a concentration of about 24×10^{13} atoms/g, using the mantle mass of 4×10^{27} g. The 'High gas' ^{40}Ar estimate is similar to that of Ballentine et al., but taking account of the amount in the continental crust (7×10^{40} atoms), to yield about 22×10^{13} atoms/g in the MORB source (Table 10.5(b)). The 'Medium gas' argon is reduced from this in proportion to the other gases.

For ^{36}Ar, the $^{40}Ar/^{36}Ar$ ratio appears to approach 30 000, but could be higher if there is atmospheric contamination, or lower if some of the spread represents real heterogeneity in the MORB source, as discussed above. Thus the ^{36}Ar abundance is not very well constrained in this interpretation (Table 10.5(b)).

10.8.4 Modelling evolution in the MORB and OIB sources

A key test of the proposed residence of mantle noble gases in hybrid pyroxenites is whether the difference in residence times between the MORB source and the D″ region can explain the differences between the noble gas signatures of MORBs and OIBs. This requires a way to calculate the evolution of the noble gases in the mantle. Such a calculation can also test the overall plausibility of the proposal in the light of results from numerical models of mantle evolution [156]. Here a simple method is presented for calculating the evolution of noble gas concentrations in the mantle, taking account of degassing and radioactive generation. This method is then applied in the next section, using the estimated ranges just discussed as boundary conditions.

We used the 'model time' of Section 10.4.3 to calculate the remaining primitive fraction of the mantle and the rate of accumulation of subducted crust in the mantle as a function of time, and later to calculate tracer residence times (Section 10.7.3). The same approach can be used to estimate rates of degassing of the mantle. By the same logic as the survival of primitive mantle, the mantle's complement of primordial volatiles will also decline exponentially with model time if the rate of degassing is constant with respect to t_m:

$$c_i = c_i^0 \exp\left(-\frac{t_m}{\tau_g}\right), \tag{10.9}$$

where c_i is the concentration of species i, c_i^0 is its initial concentration, and $\tau_g = m_i/r_i$ is the *degassing timescale*, where m_i is the mass of species i in the mantle and r_i is the rate at which it is being removed from the mantle.

More generally, the concentration of a radiogenic species will change according to

$$\frac{\Delta c_i}{\Delta t_m} = -\frac{c_i}{\tau_g} + g_i, \tag{10.10}$$

where g_i is the rate of radioactive generation of the species and τ_g need not be constant.

From the previous section, the observed degassing rate of ^{3}He is $r_3 = 1060$ mol/yr, and the conventionally inferred concentration in the mantle source is about $c_3 = 10^9$ atoms ^{3}He/g (Table 10.5(a)). The mass of ^{3}He in the mantle is $m_3 = Mc_3$, using the mass of the mantle $M = 4 \times 10^{27}$ g. The degassing timescale for ^{3}He is then $\tau_g = m_3/r_3 = 6.7$ Gyr. If the concentration of ^{3}He is two or three times higher, then m_3 is larger and the degassing timescale is correspondingly longer. Thus the degassing of the MORB source may be relatively slow.

Table 10.6. *Noble gas evolution, 'Medium gas' case*[a].

(a) Concentrations (atoms/g)

	^{4}He (10^{13})	^{3}He (10^8)	^{21}Ne (10^6)	^{22}Ne (10^8)	^{40}Ar (10^{13})	^{36}Ar (10^8)
Initial	23	383	141	43	0.007	718
MORB source, present	25	26	20	2.9	12	48
OIB source (2.5 Ga)	34	76	41	8.5	16	142
OIB source (3.75 Ga)	41	130	63	15	18	244
Production (4.5 Ga)	45	–	21	–	27	–

(b) Ratios

	^{4}He/^{3}He	^{21}Ne/^{22}Ne	^{40}Ar/^{36}Ar
MORB source	98 400	0.070	24 400
OIB source (2.5 Ga)	44 800	0.049	10 900
OIB source (3.75 Ga)	31 600	0.043	7 500

[a] $\tau_g = 10.1$ Gyr; early degassing enhanced by 3 for 0.5 Gyr; initial non-nucleogenic concentrations increased by 4.

To include radiogenic species, the uranium concentration in the MORB source is taken to be 10 ng/g, as deduced in Section 10.2. Thorium and potassium concentrations are obtained by assuming Th/U = 3.8 and K/U = 1.3×10^4 [184]. This amount of radioactivity generates 4.5×10^{14} atoms ^{4}He/g over 4.5 Gyr. ^{21}Ne production is obtained using the production ratio ^{4}He/^{21}Ne = 2.2×10^7 [230]. Total production of radiogenic species over 4.5 Gyr is included in Table 10.6(a).

10.8.5 Application of the model

Equation (10.10) can be integrated to calculate concentrations as a function of time. The strategy is to estimate initial concentrations of the non-radiogenic isotopes, to integrate forward in time, and then to test whether the radiogenic isotopes reach appropriate present values. A successful calculation turns out to require a period of enhanced degassing early in Earth history [227], as we will see, and as has been inferred previously [233]. The calculation shown here is based on the 'Medium gas' case of Table 10.5(b). It is meant to be indicative, not definitive, but is nevertheless sufficient to demonstrate the initial plausibility of the present interpretation.

Preliminary initial concentrations of the non-radiogenic isotopes were estimated from present values using Eq. (10.9). A low initial value of ^{40}Ar was assumed. The initial value of ^{4}He was obtained by assuming the ^{4}He/^{3}He ratio of 6000 measured in Jupiter's atmosphere [234]. The initial value of ^{21}Ne is obtained by assuming the solar ratio ^{21}Ne/^{22}Ne $= 0.0328$ [180].

If all of the radiogenic ^{4}He were retained in the mantle, this would imply a present ^{4}He/^{3}He ratio of about 280 000, triple the observed MORB value. Even with degassing accounted for, the calculated present ratio is still about 150 000, because of the slow degassing rates inferred above.

However, it is plausible that the degassing rate was higher during the first 0.5 Gyr or so: during this time the mantle temperature drops rapidly from a high post-accretion value to a value sustained by radiogenic heating, according to thermal evolution calculations (Chapter 9), and the rate of mantle overturn would be much higher than is implicit in Eq. (10.9). Because ^{4}He is generated at a much higher rate early on, proportionately more of it is removed by the early degassing and the effect is to reduce later values of ^{4}He/^{3}He. Argon values behave similarly.

This early degassing phase was crudely approximated in these calculations by increasing the degassing rate for the first 0.5 Gyr by a factor of 3 over that inferred from Eq. (10.9). This is sufficient to bring ^{4}He to reasonable present values. A little trial and error then established that multiplying the preliminary initial concentrations of primordial species (^{3}He, initial ^{4}He, ^{22}Ne, solar ^{21}Ne, ^{40}Ar) by a factor of 4 over those inferred from Eq. (10.9) compensated for the extra early loss of these species. The resulting initial values are included in Table 10.6(a). These values are still plausible relative to estimates of initial abundances based on solar ratios [180, 234].

The consequent evolution of helium, neon and argon isotopes in the MORB source is shown in Figure 10.25 (solid curves). The resulting present MORB-source values of all species are given in Table 10.6(a), and the corresponding ratios are given in Table 10.6(b). Good agreement with observed ratios has been achieved, and present concentrations are near the ranges given in Table 10.5(b), except for argon, which is low, and will be discussed later. The early phase of more rapid degassing is equivalent to the inference previously made that much of the mantle's argon must have degassed within the first 1 Gyr of Earth history [214, 233]. These results demonstrate that a relatively simple and plausible model based on the hybrid pyroxenite hypothesis is capable of reproducing the MORB-source noble gas observations reasonably well.

A stronger test of the hypothesis comes from calculating the OIB concentrations. Following the discussion in Section 10.8.1, in which it was noted that the OIB material has a longer residence time and consequently degasses more slowly, an OIB evolution has been calculated simply by assuming that the degassing rate is 60% of the MORB-source degassing rate, with all other parameters the same. This degassing rate would correspond to that component of OIB material with a final age of about 2.5 Ga, compared with a mean MORB age of about 1.5 Ga. (Recall that

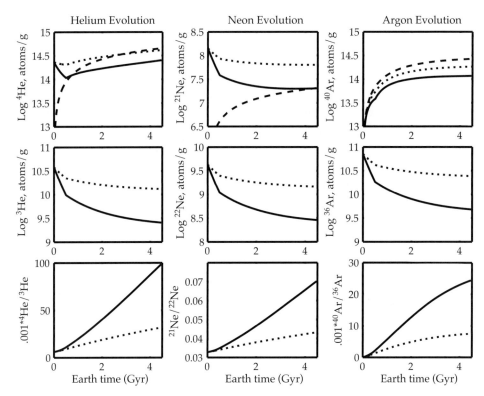

Figure 10.25. An indicative calculated evolution of He, Ne and Ar isotope concentrations and ratios for the 'Medium gas' case of Table 10.5. MORB source: solid curves; OIB source: dotted curves; radiogenic production: dashed curves. The calculation includes faster degassing in the first 0.5 Gyr (see text). From Davies [227]. Copyright American Geophysical Union.

material of essentially all ages is present in the D″ accumulations; Figure 10.23.) The resulting present values are given in Table 10.6(a) and ratios in Table 10.6(b). The ratios are representative of the mid-range of observed OIB values.

Another OIB calculation was run assuming a degassing rate 40% of the MORB rate. This would correspond to a component of OIB material with a final age of about 3.75 Ga. The resulting present values are also given in Table 10.6(a) and (b), and the evolutions are included in Figure 10.25 (dotted curves). The present ratios are representative of the most unradiogenic OIB noble gas values.

Thus the model is able to span the observed range from MORB to the most unradiogenic OIB ratios by adjusting the residence time to be consistent with results from numerical models, as hypothesised in Section 10.8.1. The proposal that noble gases are carried in a hybrid pyroxenite component of the mantle is thus able to account straightforwardly and quantitatively for He, Ne and Ar in the mantle (except for the argon abundance, to be discussed shortly). Particularly notable is the fact that the hitherto enigmatic occurrence of unradiogenic helium in OIBs has been accounted for in a straightforward way using numerical models as a guide.

10.8.6 The ^{40}Ar budget

The ^{40}Ar results require additional discussion. The ^{40}Ar abundance in the MORB source yielded by the evolution model, 12×10^{13} atoms/g, is only 55% of that required (22×10^{13} atoms/g) to balance the usual estimate of the Earth's ^{40}Ar budget (50×10^{13} atoms/g). The inferred concentration in D″ is a little higher, but D″ and its related superpiles are equivalent to only about 2% of the mass of the mantle, so it can hold only perhaps 3%. With about 50% in the atmosphere and 27% in the MORB source, this seems to leave about 20% still unaccounted for. However, this deficit is much smaller than has been conventionally inferred [225], and we should consider whether it is significant given the uncertainties involved.

One uncertainty is in the K/U ratio of the mantle, and there is some debate about whether it might be lower [235, 236] or higher [237] than the value of 13 000 [184] commonly used. Another possibility is that there is less K in the continental crust than is usually estimated. Taylor and McLennan [238] estimated the K_2O content of the continents to be 1.1%, compared with Rudnick and Fountain's 1.9% [142]. The continental crust is so heterogeneous that it is doubtful any estimates have great accuracy. If, as an illustration, we sum the mass of K in the crust using Taylor and McLennan's estimate (crustal mass 2.6×10^{25} g; K mass 2.38×10^{23} g), D″ (mass 8.5×10^{25} g; K concentration 650 μg/g and K mass 0.55×10^{23} g) and MORB source (mass 390×10^{25} g; K mass 5.2×10^{23} g), we get a total of 8.1×10^{23} g of K and a bulk silicate Earth concentration of 203 μg/g. This compares with the more usual estimate of around 240 μg/g [140], but the total K content of the Earth has significant uncertainty because of the volatility of K during the planetary accretion process (Section 10.2).

This reduced K content (203 μg/g) would yield 169×10^{40} atoms of ^{40}Ar over 4.5 Gyr. The inventory for the present model is 48×10^{40} atoms in the MORB source and 0.7×10^{40} atoms in D″. The amount in the continental crust is not well determined but estimated to be about 7×10^{40}, and the amount in the atmosphere is 99×10^{40} [225, 229]. These sum to 155×10^{40} atoms. An evolution based on the 'High gas' case of Table 10.5 yields a final total of 170×10^{40} atoms, so between them the models span the bulk silicate amount just estimated. Thus a lower crustal abundance of K, and a consequent lower global abundance, is consistent with the present models.

Finally, we should not really assume that the Earth has retained all of its volatiles, because some of the atmosphere may have been blasted off by very large impacts during the late stages of accretion, or by the late heavy bombardment if that was a separate episode [239]. Bombardment persisted until about 3.8 Ga, by which time about one-third of the Earth's ^{40}Ar had been generated. If mantle degassing was rapid during the first 0.5 Gyr or so, then it is conceivable that the order of 10% of the Earth's ^{40}Ar was removed from the Earth entirely [240].

Thus the ^{40}Ar constraint is not as stringent as has been claimed. There are uncertainties in the amount of potassium in the continental crust and in the bulk silicate Earth, and there may have been some early loss of ^{40}Ar from the atmosphere. The present mantle model may account for the total ^{40}Ar budget of the Earth when these uncertainties are allowed for.

10.8.7 Alternative interpretations

The presence of ^{3}He in some samples was never a reason in itself for inferring a primitive source. Certainly ^{3}He is not produced in the Earth, so any that remains is 'primitive', but it is hardly surprising that some small amount is still trickling out of the Earth. The real question is how the present amount compares with plausible initial amounts, and with the amount of ^{4}He produced in the meantime, questions addressed implicitly in the interpretation just presented. The jump to presuming an 'undegassed' mantle reservoir, meaning one that has retained all radiogenic gases produced over 4.5 Gyr, was evidently conditioned by the presumption of a 'primitive' lower mantle, originally proposed to explain a handful of dubious Nd data. That presumption has failed to survive critical scrutiny, as we have seen.

A recent proposal by Tolstikhin and Hofmann [241] is not so readily dismissed. They proposed that D″ was formed very early in Earth history from a foundered mixture of late-accreting material and early basaltic crust. The accreting material is presumed to carry high concentrations of solar noble gases and excess iron. The excess iron would increase the density and thus tend to stabilise the layer and allow it to survive to the present. This model deals better with the main constraints, though it requires rather detailed specification of the initial state of D″ and may be difficult to test. Nevertheless, it has more merit than many predecessors and deserves to be critically evaluated.

Albarède [235] suggested that unradiogenic helium comes from sources with low U, rather than high He, and that such a source would be produced if He was less incompatible than U during melting. However, the source would then be a melt residue that would be refractory and have very low concentrations of both elements, so the helium would not be readily evident in subsequent melting events. The residue would also tend to be less dense than average mantle, so it is not obvious why it would show up preferentially in mantle plumes. Some other hypotheses that have been proposed at various times are discussed by Davies [227].

10.9 Assessment of mantle chemistry

This discussion of mantle chemistry has addressed the question of how it can fit into the picture of mantle dynamics developed earlier in the book. An understanding of how the refractory incompatible elements fit into mantle dynamics has emerged over the past decade or so, following the seminal work by Christensen and Hofmann [122]. Two basic questions remained unresolved, namely how the global budgets of refractory incompatible elements could be balanced in such a picture and how the noble gases might fit in. Consideration of the important role of major element heterogeneity, especially in controlling melting and melt extraction and in inhibiting chemical equilibration, has led to plausible resolutions of both questions. These proposals will, of course, need to be debated and tested.

The interpretations presented here have been quantified to varying degrees. The quantification of refractory incompatible element behaviour is at the level of numerical models of thermochemical mantle dynamics. The quantification of the role of noble gases is quite simplified at this stage, but any quantification that achieves plausible concordance with helium observations is an advance. It is thus reasonable to say that quantitative dynamical treatments of mantle geochemistry are becoming established.

Earlier interpretations invoking a primitive or undegassed reservoir were never soundly based even on geochemical grounds, as they were contradicted by lead isotope data from the beginning and they could only explain the diversity of OIB data with additional *ad hoc* assumptions. However understandable the original presumptions might have been, in the context of some geophysicists' early presumption of an inactive lower mantle or, later, a layered mantle, the subsequent failure to critically examine basic assumptions and the reliability of key observations, and the propensity to ignore contrary observations such as the lead isotope data, has impeded the subject for decades.

11

Assimilating mantle convection into geology

The goal of this book has been to present mantle convection as a topic accessible to all geoscientists so that it can become a routine part of their thinking. This is necessary, because it is so fundamental to so many geological processes. It is also possible, as I hope this presentation has demonstrated.

The understanding of mantle convection that has been developed in the book is fairly well established. There remains some debate about a possible layer in the deep mantle, apart from D'', though the debates usually seem to be conducted in ignorance of the heat flow constraint given in Chapter 8. There is also some debate about the existence of mantle plumes, though this persists mainly in circles that are apparently ignorant of the physics of mantle upwellings given in Chapter 7. In any case readers should be able to follow the essence of such continuing debates, which are a normal part of any scientific topic. Hopefully they will be armed with a better-informed perspective.

Two kinds of implication have also been discussed in this book: for the tectonic evolution of the Earth, and for the chemistry of the mantle. The early tectonic evolution of the Earth is still unclear. Observational constraints are so sparse and possibilities still so diverse that progress will require a continuing conversation among all those involved, and a grasp of the essence of mantle evolution models will be important. Although this topic is still rudimentary in some respects, it is notable how much more clearly defined it is now compared with, say, two decades ago.

After a long period of debate and confusion, mantle geochemistry now seems to be being brought into the fold of mantle dynamics. Hopefully this will focus debates in more productive directions. Major element heterogeneity emerges as a particularly important aspect. It leads to much more complicated behaviour in melting and melt migration, and progress will therefore be more difficult, but some useful ideas have already emerged from experimental and observational work of the past decade or so. The presentation here has included recent work at the forefront of research, so it is less well established than the basics of mantle convection, but those concerned with it hopefully will be equipped with a clearer appreciation of the nature, role and importance of mantle dynamics as the debates are carried forward.

Appendix A

Exponential growth and decay

At a number of places in this book, a situation has given rise to exponential growth or decay. Although these situations have been analysed without using calculus, as promised, it may be useful to those who know some basic calculus to present the more rigorous solution that calculus allows. Exponential behaviour arises in a standard way, so we can start with a general situation. The resulting solution can then be adapted to particular situations by appropriately identifying the variables. Some of the particular situations will be covered in this appendix, whereas others will be covered in later appendices.

A.1 Exponential solution

Suppose something is growing larger, and the rate at which it grows is proportional to its present size. Let's call the something y and its rate of growth v. Then our situation is described by the relationship

$$v = ay. \tag{A.1}$$

But if v is the rate of change of y, we can also write

$$v = \frac{dy}{dt},$$

so Eq. (A.1) becomes

$$\frac{dy}{dt} = ay. \tag{A.2}$$

To solve this equation, we can treat the differentials as infinitesimal increments and rearrange the equation as

$$\frac{\delta y}{y} = a\delta t. \tag{A.3}$$

Each side is now in a form whose integral is known:

$$\int \frac{\delta y}{y} = \ln(y),$$

where ln is the natural logarithm, and

$$\int a\delta t = at.$$

The integrations also yield constants of integration that can be combined into one, so Eq. (A.3) becomes

$$\ln(y) = at + c.$$

Taking the exponential of both sides, this becomes

$$y = \exp(at + c) = \exp(at)\exp(c). \tag{A.4}$$

To evaluate the constant, we need to know a value of y at a particular time, t. Suppose that, at time $t = 0$, y has the value y_0. Putting these values into Eq. (A.4) gives

$$y_0 = \exp(0)\exp(c) = \exp(c).$$

So Eq. (A.4) becomes

$$y = y_0 \exp(at). \tag{A.5}$$

This is the basic equation of exponential growth.

If it should happen that y is *decreasing* at a rate proportional to its size, the situation can be covered by taking a to be negative. For example, sometimes a takes the form

$$a = -1/\tau.$$

The logic of solution follows exactly as that just described, and the solution takes the form

$$y = y_0 \exp(-t/\tau). \tag{A.6}$$

This is the basic equation of exponential decay, including radioactive decay.

We can also define a as $1/\tau$ in Eq. (A.5). Then τ is a growth (or decay) time constant, an 'e-folding' time. In other words, with each elapse of time $t = \tau$, y increases by the factor $e = 2.178$. It is related to a half-life or a doubling time (depending on whether it is decreasing or increasing). From Eq. (A.5), we can deduce that y will be twice y_0 when $t = 0.693\tau = \ln(2)\,\tau$, where ln denotes a natural logarithm. Thus we can say that the doubling time is $\tau_2 = \ln(2)\,\tau$. Similarly the half-life of decay in Eq. (A.6) is $\tau_{1/2} = \ln(2)\,\tau$.

A.2 Post-glacial rebound

In Section 4.2, a relationship was derived between the depth, d, of a depression caused by pre-existing ice and the rate of uplift, v, of the floor of the depression. Equation (4.8) expresses the relationship as

$$\mu = g(\rho_m - \rho_w)dR/v,$$

which we can rearrange as

$$v = g(\rho_m - \rho_w)Rd/\mu.$$

However, v is just the rate of decrease of d, so we can write

$$\frac{\partial d}{\partial t} = -\frac{gR\Delta\rho}{\mu}d, \qquad (A.7)$$

where the differentials have been written with the ∂ symbol to distinguish them from the variable d, and $\Delta\rho = (\rho_m - \rho_w)$.

The collection of constants on the right-hand side of Eq. (A.7) has the dimensions of inverse time, so let's define a time constant as

$$\tau = \mu/gR\Delta\rho. \qquad (A.8)$$

Then Eq. (A.7) can be written as

$$\frac{\partial d}{\partial t} = -\frac{d}{\tau}. \qquad (A.9)$$

This has the same form as Eq. (A.2). We can identify d with y and a with $-1/\tau$ and take the solution from Eq. (A.5):

$$d = d_0 \exp(-t/\tau). \qquad (A.10)$$

Thus we derive that the depth of the post-glacial depression decreases exponentially with time. As Figure 4.2 shows, an exponential decline closely fits the observed change in relative sea level in Fennoscandia, and the best-fit exponential has a decay timescale of 4.6 kyr. If we identify this with τ in Eq. (A.8), then we get another expression for mantle viscosity:

$$\mu = \tau gR\Delta\rho. \qquad (A.11)$$

Using $g = 10 \, \text{m/s}^2$, $R = 1000 \, \text{km}$ and $\Delta\rho = 2300 \, \text{kg/m}^3$, this yields $\mu = 3.3 \times 10^{21}$ Pa s.

A.3 Rayleigh–Taylor instability

In Section 7.3.2, the instability of a fluid layer is analysed by considering a bulge of height h on an interface between low-density fluid below and higher-density fluid above. Equation (7.13) is given for the velocity, v, of the top of the bulge:

$$v = (g\Delta\rho w/\mu)h.$$

The form of this relationship is very similar to the post-glacial relationship in the preceding section, except here the bulge is growing, so $v = dh/dt$ and

$$\frac{\partial h}{\partial t} = \frac{gw\Delta\rho}{\mu}h = \frac{h}{\tau}. \qquad (A.12)$$

This has the same form as Eq. (A.2), so we can identify h with y and a with $1/\tau$ and write the solution as

$$h = h_0 \exp(t/\tau), \qquad (A.13)$$

which is the same as Eq. (7.14). This describes exponential growth of the height of the bulge. In other words, the bulge is unstable and grows with a timescale of τ.

Appendix B

Thermal evolution details

B.1 Heat generation

The heat generation per unit mass due to radioactive heating is given by

$$H_R = \mathrm{Ur} U_e \sum_{i=1}^{4} h_i \exp[\lambda_i(t_E - t)], \tag{B.1}$$

where Ur is the Urey ratio, defined by Eq. (9.5), U_e is the equivalent uranium concentration (38 ng/g) that would yield the observed heat loss, the index i refers to the isotopes ^{238}U, ^{235}U, ^{232}Th and ^{40}K, h_i is the heat production per unit mass of uranium, λ_i is the decay constant and t_E is the age of the Earth (4.5 Ga). Values for the required parameters are given in Table B.1.

B.2 Parametrised thermal evolution model

The thermal evolution model in Figure 9.1 is a variation on the models reported in Davies [65]. These models include the effects of the core crystallising to form the inner core. There are many details of this process, and many parameters involved, which won't be reproduced here, as they can be found in that paper. The other main parameters and outputs of this model are given in Table B.2.

There is one feature of this solution worth noting, as it adds usefully to the examples in Davies [65]. In the present calculation, the starting temperature of the core was increased from 4000 °C to 4500 °C, relative to Case 8 of the paper. This increased the core heat loss to within the range estimated in Chapter 7. It also increased the energy available within the core to drive the magnetic dynamo. There are several apparently conflicting constraints that the paper attempted to reconcile, one of them being the need for enough energy to drive the dynamo, which is not well constrained but was estimated at the time to be around 1 TW. Core energy dissipation and the inner core growth are included in Figure B.1. The thermal dissipation starts above 1 TW and decreases, but after the inner core starts to crystallise the dissipation driven by compositional convection adds to this and boosts it back above 1 TW. The lowest total dissipation is 0.51 TW at the time the inner core nucleates. This calculation more easily reconciles the various constraints than those given in the paper above [65].

Table B.1. *Parameters of heat-producing isotopes.*

Isotope, i	Half-life (Ga)	Decay const.[a], λ_i (Ga^{-1})	Power[a] (μW/kg element)	Element/U[b] (g/g)	Power, h_i (μW/kg U)
^{238}U	4.468	0.155	94.35	1	94.35
^{235}U	0.7038	0.985	4.05	1	4.05
^{232}Th	14.01	0.049	26.6	3.8	101.1
^{40}K	1.250	0.554	0.0035	1.3×10^4	45.5
				Total	245

[a] Stacey [141].
[b] Galer et al. [242].

Table B.2. *Parameters of the thermal evolution model of Figure 9.1 (except for parameters related to crystallisation of the inner core).*

Quantity	Symbol	Value
Inputs:		
Urey ratio	Ur	0.69
Plate adjustment factor		0.16
Plume adjustment factor		0.12
Reference mantle temperature	T_r	1300 °C
Reference mantle viscosity	μ_r	10^{21} Pa s
Viscous activation energy	E^*	325 kJ/mol
Initial temperature, upper mantle	T_U	3500 °C
Initial temperature, lower mantle	T_L	4500 °C
Crystallisation temperature (centre)		6070 °C
Outputs:		
Final temperature, upper mantle	T_U	1306 °C
Final temperature, lower mantle	T_L	2306 °C
Final temperature, outer core	T_C	3876 °C
Final surface heat loss	Q_S	36.4 TW
Final core heat loss	Q_C	7.6 TW
Final radiogenic heating	Q_R	24.6 TW
Maximum thermal dissipation		1.23 TW
Final thermal dissipation		0.41 TW
Final total dissipation		1.82 TW

Such a high starting temperature for the core gains some rationale from recent suggestions that the core might be superheated. The idea is that a large amount of heating occurs as core material separates from the mantle (enough to heat the whole Earth by around 2000 °C), but that most of the heating occurs in the core material. The core material might have to collect into large bodies before it can sink through the relatively cool and stiff mantle, in which case it probably would not thermally equilibrate with the

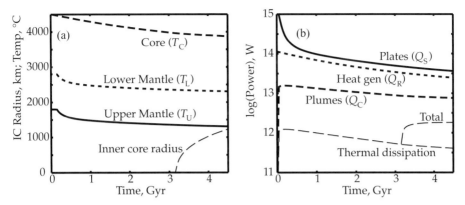

Figure B.1. The thermal evolution of Figure 9.1 including core-related results (long-dashed, thin lines). (a) The radius of the inner core. (b) Thermal dissipation in the core (lower curve) and total dissipation (upper curve) including dissipation from compositional convection driven by crystallisation of the inner core.

mantle material. The result would be a core considerably hotter than the mantle, even at the core–mantle boundary. In the present model, the lower mantle starts at the same temperature as the core, but the mantle cools very quickly, so a large temperature difference is soon established.

B.3 Numerical thermal evolution model

The results shown in Figures 9.2 and 9.3 are Case 5 of Davies [156], where full details can be found. The most relevant inputs and outputs are shown in Table B.3. This model is not tailored as closely to fit observed heat flows, and they are a little higher than either the model of Figure 9.4 or the observed values. Nevertheless, the model demonstrates basically similar behaviour of the numerical and parametrised models.

B.4 Basalt tracers

Material of basaltic composition is represented in the numerical convection models by tracers. The tracers each carry a small mass anomaly, corresponding to the different density of the basaltic component at depth relative to average mantle. Tracers are advected with the mantle flow using a fourth-order Runge–Kutta algorithm [139, 156, 203].

As fluid rises through the depth of the solidus, melting is simulated by removing tracers and placing them in a thin crustal layer. This simulates pressure release melting, and leaves a layer of fluid depleted of tracers, simulating the depleted residue of the oceanic crust source. The depleted layer is less dense than surrounding material containing heavy tracers.

To calculate the depth of the solidus, the mantle is presumed to be heterogeneous in major element composition. Since eclogite has a lower solidus temperature than peridotite, the depth at which first melting occurs will be controlled by the eclogite. The

Table B.3. *Parameters of the numerical thermal evolution model of Figure 9.2.*

Quantity	Symbol	Value
Inputs:		
Reference mantle temperature	T_r	1300 °C
Reference upper-mantle viscosity	μ_r	3×10^{20} Pa s
Reference lower-mantle viscosity		10^{22} Pa s
Viscous activation energy	E^*	325 kJ/mol
Initial temperature		1650 °C
Initially estimated Urey ratio	Ur	0.71
Outputs:		
Final temperature, upper mantle	T_U	1284 °C
Final temperature, mean mantle	T	1269 °C
Final surface heat loss	Q_S	40.1 TW
Final radiogenic heating	Q_R	35.7 TW
Final Urey ratio	Ur	0.89

melting depth in the models is estimated from the mean upper-mantle temperature and an approximation to the eclogite solidus versus depth of Yasuda *et al.* [80]:

$$z_\text{melt} = 0.478 \left(T_\text{up} - 1020\right). \tag{B.2}$$

The variation of crustal thickness with temperature in undepleted mantle is taken from McKenzie and Bickle's estimates [243] for fertile (undepleted) mantle, calibrated to yield 7 km at 1280 °C:

$$d_\text{crust} = 7 + 0.085 \left(T_\text{up} - 1280\right). \tag{B.3}$$

Units in both formulae are km and °C. The method of estimation of the crustal thickness is not strictly consistent with the eclogite melting implied by the melting model, but it is assumed to usefully represent the trend for this exploration.

Subduction returns the crustal layer and the depleted layer to the mantle, where they are stirred by mantle convection. These and other features of the model are described in more detail by Davies [203]. The simulated melting is suppressed under the 'continental' region in the centre, since the upper boundary there is taken to be the base of continental lithosphere at around 200 km depth, under which little melting should occur.

Because heavy tracers tend to accumulate at the bottom of the model, it is necessary to ensure that their concentration does not exceed that corrresponding to pure basaltic component. This is done during the tracer advection process by checking for each tracer whether the limit has already been exceeded for the local grid cell. If so, the tracer is moved up until it reaches a cell whose concentration is below the limit.

B.5 Models with basalt tracers

The results shown in Figure 9.5 are from an unpublished model, and those in Figure 9.10 are Case 2 of Davies [156], where full details can be found. Figure 9.5 is similar to Case 1

Table B.4. *Parameters of the numerical thermal evolution model of Figures 9.5 and 9.10.*

Quantity	Figure 9.5	Figure 9.10
Final temperature, upper mantle, T_U	1285 °C	1274 °C
Final temperature, mean mantle, T	1276 °C	1302 °C
Final surface heat loss, Q_S	41.3 TW	39.0 TW
Final radiogenic heating, Q_R	35.7 TW	35.7 TW
Final Urey ratio, Ur	0.86	0.92

except that there is no barrier to flow at 660 km depth due to phase transformations; in other words the Clapeyron slope is zero. General inputs are the same as those given for the model of Figure 9.2 (Section B.3). The most relevant outputs are shown in Table B.4.

Appendix C
Chemical evolution details

C.1 Fraction of primitive mantle

We can find the solution of Eq. (10.4) by writing it in terms of differentials:

$$\frac{dp}{dt} = -\frac{p}{\tau}.$$

This has the form of Eq. (A.2) with a identified as $-1/\tau$, so its solution is, from Appendix A, Eq. (A.6)

$$p = p_0 \exp(-t/\tau). \tag{C.1}$$

At time $t = 0$, $p = 1$, so $p_0 = 1$, which then yields Eq. (10.5).

C.2 Fraction of crust in the mantle

Following the discussion in Section 10.4.4, the net amount of oceanic crust that accumulates in the mantle during a time interval Δt will be

$$\Delta m_c = (\phi_c - f\phi)\Delta t.$$

Then f will increase by $\Delta m_c/M$, and the rate of increase of f will be

$$\frac{\Delta f}{\Delta t} = \frac{1}{M}(\phi_c - f\phi). \tag{C.2}$$

As discussed in Section 10.4.4, we expect f to approach a maximum level, at which the rate of addition, ϕ_c, balances the rate of removal, $f\phi$. This condition will define the maximum level, f_m, or in other words

$$f_m = \phi_c/\phi. \tag{C.3}$$

We can substitute $\phi_c = f_m \phi$ in Eq. (C.2) and get

$$\frac{\Delta f}{\Delta t} = \frac{\phi}{M}(f_m - f) = \frac{1}{\tau}(f_m - f). \tag{C.4}$$

This equation says that the rate of increase of f gets smaller as f approaches the maximum, f_m. That fits the qualitative understanding developed in Section 10.4.4.

C.2 Fraction of crust in the mantle

To solve this equation, it is helpful to define $w = f_m - f$. Then, using differentials, $dw/dt = -df/dt$ and Eq. (C.4) becomes

$$\frac{dw}{dt} = -\frac{w}{\tau}.$$

This is the classic equation describing exponential decay, with the solution (Section A.1)

$$w = w_0 \exp(-t/\tau).$$

When $t = 0, f = 0$, so $w_0 = f_m$. Now substituting the definition of w, we get

$$f_m - f = f_m \exp(-t/\tau)$$

or

$$f = f_m \left[1 - \exp(-t/\tau)\right], \tag{C.5}$$

which is Eq. (10.7).

References

1. Davies, G.F., *Dynamic Earth: Plates, Plumes and Mantle Convection.* 1999, Cambridge: Cambridge University Press. 460p.
2. Kennett, B.L.N., E.R. Engdahl, and R. Buland, Constraints on seismic velocities in the earth from travel times. *Geophys. J. Int.*, 1995. **122**: p. 108–124.
3. Montagner, J.P. and B.L.N. Kennett, How to reconcile body-wave and normal-mode reference Earth models? *Geophys. J. Int.*, 1996. **125**: p. 229–248.
4. Davies, G.F., Mantle plumes, mantle stirring and hotspot chemistry. *Earth Planet. Sci. Lett.*, 1990. **99**: p. 94–109.
5. Mohorovičić, A., Das Beben vom 8.X.1909. *Jahrb. Met. Obs. Zagreb (Agram.)*, 1909. **9**: p. 1–63.
6. Hess, H.H., History of ocean basins, in *Petrologic Studies: a Volume in Honor of A.F. Buddington*, A.E.J. Engel, H.L. James, and B.F. Leonard, Editors. 1962, Boulder, CO: Geological Society of America. p. 599–620.
7. Menard, H.W., *The Ocean of Truth.* 1986, Princeton, NJ: Princeton University Press. 353p.
8. Barrell, J., The strength of the earth's crust. *J. Geol.*, 1914. **22**: p. 655–683.
9. ETOPO5 (Topography of the Earth, 5 minute grid), National Geophysical Data Center, US National Oceanic and Atmospheric Administration, Boulder, CO.
10. Marty, J.C. and A. Cazenave, Regional variations in subsidence rate of oceanic plates: a global analysis. *Earth Planet. Sci. Let.*, 1989. **94**: p. 301–315.
11. Sclater, J.G., C. Jaupart, and D. Galson, The heat flow through the oceanic and continental crust and the heat loss of the earth. *Rev. Geophys.*, 1980. **18**: p. 269–312.
12. Glen, W., *Continental Drift and Plate Tectonics.* 1975, Columbus, OH: Charles E. Merrill. 188p.
13. Hallam, A., *A Revolution in the Earth Sciences.* 1973, Oxford: Clarendon Press. 127p.
14. Wegener, A., *Die Entstehung der Kontinente und Ozeane.* 1st edn. 1915, Brunswick: Vieweg.
15. Jeffreys, H., *The Earth, its Origin, History and Physical Constitution.* 6th edn. 1976, Cambridge: Cambridge University Press.
16. Holmes, A., *Principles of Physical Geology.* 1st edn. 1944, Edinburgh: Thomas Nelson.
17. Daly, R.A., *Strength and Structure of the Earth.* 1940, New York: Prentice-Hall. Facsimile edition. 1969, New York: Hafner. 434p.
18. du Toit, A.L., *Our Wandering Continents.* 1937, Edinburgh: Oliver and Boyd.

19. Carey, S.W., The tectonic approach to continental drift, in *Continental Drift; a Symposium*, S.W. Carey, Editor. 1958, Hobart: University of Tasmania, Geology Department. p. 177–358.
20. Runcorn, S.K., Paleomagnetic evidence for continental drift and its geophysical cause, in *Continental Drift*, S.K. Runcorn, Editor. 1962, New York: Academic Press. p. 1–40.
21. Dietz, R.S., Continent and ocean evolution by spreading of the sea floor. *Nature*, 1961. **190**: p. 854–857.
22. Heezen, B.C., The rift in the ocean floor. *Sci. Am.*, 1960. **203**: p. 98–110.
23. Wilson, J.T., A new class of faults and their bearing on continental drift. *Nature*, 1965. **207**: p. 343–347.
24. Wilson, J.T., Evidence from islands on the spreading of the ocean floor. *Nature*, 1963. **197**: p. 536–538.
25. Wilson, J.T., A possible origin of the Hawaiian islands. *Can. J. Phys.*, 1963. **41**: p. 863–870.
26. McDougall, I., Age of shield-building volcanism of Kauai and linear migration of volcanism in the Hawaiian Island chain. *Earth Planet. Sci. Lett.*, 1979. **46**: p. 31–42.
27. McDougall, I. and D.H. Tarling, Dating of polarity zones in the Hawaiian islands. *Nature*, 1963. **200**: p. 54–56.
28. Wilson, J.T., Continental drift. *Sci. Am.*, 1963. **208** (April): p. 86–100.
29. Bucher, W.H., *The Deformation of the Earth's Crust*. 1933, Princeton: Princeton University Press. Facsimile edition. 1957, New York: Hafner. 518p.
30. Heirtzler, J.R., *et al.*, Marine magnetic anomalies, geomagnetic field reversals, and motions of the ocean floor and continents. *J. Geophys. Res.*, 1968. **73**: p. 2119–2136.
31. Sykes, L.R., Seismicity of the South Pacific Ocean. *J. Geophys. Res.*, 1963. **68**: p. 5999–6006.
32. Sykes, L.R., The seismicity of the Arctic. *Bull. Seismol. Soc. Am.*, 1965. **55**: p. 501–518.
33. Sykes, L.R., Mechanism of earthquakes and nature of faulting on the mid-ocean ridges. *J. Geophys. Res.*, 1967. **72**: p. 2131–2153.
34. Maxwell, A.E., *et al.*, Deep sea drilling in the South Atlantic. *Science*, 1970. **168**: p. 1047–1059.
35. Airy, G.B., *Phil. Trans. R. Soc. Lond.*, 1855. 145: p. 101–104.
36. Hall, J., *Geology of New York State*. 1859. p. 69.
37. Mitrovica, J.X., Haskell [1935] revisited. *J. Geophys. Res.*, 1996. **101**: p. 555–569.
38. Haskell, N.A., The viscosity of the asthenosphere. *Am. J. Sci.*, ser. 5, 1937. **33**: p. 22–28.
39. Kohlstedt, D.L., B. Evans, and S.J. Mackwell, Strength of the lithosphere: constraints imposed by laboratory experiments. *J. Geophys. Res.*, 1995. **100**: p. 17587–17602.
40. Tozer, D.C., Heat transfer and convection currents. *Phil. Trans. R. Soc. Lond. A*, 1965. **258**: p. 252–271.
41. Mitrovica, J.X. and A.M. Forte, Radial profile of mantle viscosity: results from the joint inversion of convection and postglacial rebound observables. *J. Geophys. Res.*, 1997. **102**: p. 2751–2769.
42. Parsons, B., Causes and consequences of the relation between area and age of the ocean floor. *J. Geophys. Res.*, 1982. **87**: p. 289–302.
43. Morgan, W.J., Convection plumes in the lower mantle. *Nature*, 1971. **230**: p. 42–43.

44. Morgan, W.J., Plate motions and deep mantle convection. *Mem. Geol. Soc. Am.*, 1972. **132**: p. 7–22.
45. Morgan, W.J., Rises, trenches, great faults and crustal blocks. *J. Geophys. Res.*, 1968. **73**: p. 1959–1982.
46. Goldreich, P. and A. Toomre, Some remarks on polar wandering. *J. Geophys. Res.*, 1969. **74**: p. 2555–2567.
47. Gordon, R.G., B.C. Horner-Johnson, and R.R. Kumar, Latitudinal shift of the Hawaiian hotspot: motion relative to other hotspots or motion of all hotspots in unison relative to the spin axis (i.e. true polar wander)? *Geophys. Res. Abstr.*, 2005. **7**: p. 10233.
48. Tarduno, J.A., *et al.*, The Emperor Seamounts: southward motion of the Hawaiian hotspot plume in Earth's mantle. *Science*, 2003. **301**: p. 1064–1069.
49. Crough, S.T. and D.M. Jurdy, Subducted lithosphere, hotspots and the geoid. *Earth Planet. Sci. Lett.*, 1980. **48**: p. 15–22.
50. Duncan, R.A. and M.A. Richards, Hotspots, mantle plumes, flood basalts, and true polar wander. *Rev. Geophys.*, 1991. **29**: p. 31–50.
51. Lay, T., J.W. Hernlund, and B.A. Buffett, Core–mantle boundary heat flow. *Nature Geosci.*, 2008. **1**: p. 25–32.
52. Watts, A.B. and U.S. ten Brink, Crustal structure, flexure and subsidence history of the Hawaiian Islands. *J. Geophys. Res.*, 1989. **94**: p. 10473–10500.
53. Turcotte, D.L. and G. Schubert, *Geodynamics: Applications of Continuum Physics to Geological Problems*. 2nd edn. 2001, Cambridge: Cambridge University Press. 528p.
54. Davies, G.F., Ocean bathymetry and mantle convection, 1. Large-scale flow and hotspots. *J. Geophys. Res.*, 1988. **93**: p. 10467–10480.
55. Sleep, N.H., Hotspots and mantle plumes: some phenomenology. *J. Geophys. Res.*, 1990. **95**: p. 6715–6736.
56. Campbell, I.H. and R.W. Griffiths, Implications of mantle plume structure for the evolution of flood basalts. *Earth Planet. Sci. Lett.*, 1990. **99**: p. 79–83.
57. Clague, D.A. and G.B. Dalrymple, Tectonics, geochronology and origin of the Hawaiian–Emperor volcanic chain, in *The Eastern Pacific Ocean and Hawaii*, E.L. Winterer, D.M. Hussong, and R.W. Decker, Editors. 1989, Boulder, CO: Geological Society of America. p. 188–217.
58. Morgan, J.P., W.J. Morgan, and E. Price, Hotspot melting generates both hotspot swell volcanism and a hotspot swell? *J. Geophys. Res.*, 1995. **100**: p. 8045–8062.
59. Wessel, P., A re-examination of the flexural deformation beneath the Hawaiian islands. *J. Geophys. Res.*, 1993. **98**: p. 12177–12190.
60. Hofmann, A.W., Sampling mantle heterogeneity through oceanic basalts: isotopes and trace elements, in *Treatise on Geochemistry*, Vol. 2: *The Mantle and Core*, R.W. Carlson, Editor. 2003, Oxford: Elsevier-Pergamon. p. 1–44.
61. Hill, R.I., *et al.*, Mantle plumes and continental tectonics. *Science*, 1992. **256**: p. 186–193.
62. Bunge, H.-P., Low plume excess temperature and high core heat flux inferred from non-adiabatic geotherms in internally heated mantle circulation models. *Phys. Earth Planet. Inter.*, 2005. **153**: p. 3–10.
63. Labrosse, S., Hotspots, mantle plumes and core heat loss. *Earth Planet. Sci. Lett.*, 2002. **199**: p. 147–56.
64. Zhong, S., Constraints on thermochemical convection of the mantle from plume heat flux, plume excess temperature, and upper mantle temperature. *J. Geophys. Res.*, 2006. **111**: B04409, doi:10.1029/2005JB003972.

65. Davies, G.F., Mantle regulation of core cooling: a geodynamo without core radioactivity? *Phys. Earth Planet. Inter.*, 2007. **160**: p. 215–229.
66. Nimmo, F., et al., The influence of potassium on core and geodynamo evolution. *Geophys. J. Int.*, 2004. **156**: p. 363–376.
67. Whitehead, J.A. and D.S. Luther, Dynamics of laboratory diapir and plume models. *J. Geophys. Res.*, 1975. **80**: p. 705–717.
68. Griffiths, R.W. and I.H. Campbell, Stirring and structure in mantle plumes. *Earth Planet. Sci. Lett.*, 1990. **99**: p. 66–78.
69. Morgan, W.J., Hotspot tracks and the opening of the Atlantic and Indian Oceans, in *The Sea*, C. Emiliani, Editor. 1981, New York: Wiley. p. 443–487.
70. Coffin, M.F. and O. Eldholm, Large igneous provinces: crustal structure, dimensions and external consequences. *Rev. Geophys.*, 1994. **32**: p. 1–36.
71. Richards, M.A., R.A. Duncan, and V.E. Courtillot, Flood basalts and hot-spot tracks: plume heads and tails. *Science*, 1989. **246**: p. 103–107.
72. White, R. and D. McKenzie, Magmatism at rift zones: the generation of volcanic continental margins and flood basalts. *J. Geophys. Res.*, 1989. **94**: p. 7685–7730.
73. Campbell, I.H., M.J. Cordery, and G. Davies. The relationship between mantle plumes and continental flood basalts. in *Proceedings of the International Field Conference and Symposium on Petrology and Metallogeny of Volcanic and Intrusive Rocks of the Midcontinent Rift System*. 1995.
74. Hofmann, A.W. and W.M. White, Mantle plumes from ancient oceanic crust. *Earth Planet. Sci. Lett.*, 1982. **57**: p. 421–436.
75. Cordery, M.J., G.F. Davies, and I.H. Campbell, Genesis of flood basalts from eclogite-bearing mantle plumes. *J. Geophys. Res.*, 1997. **102**: p. 20179–20197.
76. Leitch, A.M. and G.F. Davies, Mantle plumes and flood basalts: enhanced melting from plume ascent and an eclogite component. *J. Geophys. Res.*, 2001. **106**: p. 2047–2059.
77. Leitch, A.M., G.F. Davies, and M. Wells, A plume head melting under a rifting margin. *Earth Planet. Sci. Lett.*, 1998. **161**: p. 161–177.
78. Clouard, V. and A. Bonneville, How many Pacific hotspots are fed by deep-mantle plumes? *Geology*, 2001. **29**: p. 695–698.
79. Natland, J.H. and E.L. Winterer, Fissure control on volcanic action in the Pacific, in *Plumes, Plates and Paradigms*, G.R. Foulger, et al., Editors. 2005, Boulder, CO: Geological Society of America.
80. Yasuda, A., T. Fujii, and K. Kurita, Melting phase relations of an anhydrous mid-ocean ridge basalt from 3 to 20 GPa: implications for the behavior of subducted oceanic crust in the mantle. *J. Geophys. Res.*, 1994. **99**: p. 9401–9414.
81. Lin, S. and P.E. van Keken, Dynamics of thermochemical plumes: 2. Complexity of plume structures and its implications for mapping mantle plumes. *Geochem. Geophys. Geosyst.*, 2006. **7**(3): Q03003, doi:10.1029/2005GC001072.
82. Lin, S.-C. and P.E. van Keken, Multiple volcanic episodes of flood basalts caused by thermochemical mantle plumes. *Nature*, 2005. **436**: p. 250–252.
83. Kumagai, I., et al., Mantle plumes: thin, fat, successful, or failing? Constraints to explain hot spot volcanism through time and space. *Geophys. Res. Lett.*, 2008. **35**: L16301, doi:10.1029/2008GL035079.
84. Farnetani, C.G. and H. Samuel, Beyond the thermal plume paradigm. *Geophys. Res. Lett.*, 2005. **32**: L07311, doi:10.1029/2005GL022360.
85. Stefanick, M. and D.M. Jurdy, The distribution of hot spots. *J. Geophys. Res.*, 1984. **89**: p. 9919–9925.

86. Kerr, R.C. and C. Mériaux, Structure and dynamics of sheared mantle plumes. *Geochem. Geophys. Geosyst.*, 2004. **5**: Q12009, doi:10.1029/2004GC000749.
87. Richards, M.A. and D.C. Engebretson, Large-scale mantle convection and the history of subduction. *Nature*, 1992. **355**: p. 437–440.
88. Stacey, F.D. and D.E. Loper, Thermal histories of the core and mantle. *Phys. Earth Planet. Inter.*, 1984. **36**: p. 99–115.
89. Von Herzen, R.P., *et al.*, Heat flow and thermal origin of hotspot swells: the Hawaiian swell revisited. *J. Geophys. Res.*, 1989. **94**: p. 13783–13799.
90. Grand, S., R.D. Van Der Hilst, and S. Widiyantoro, Global seismic tomography: a snapshot of convection in the earth. *Geol. Soc. Am. Today*, 1997. **7**(4): p. 1–7.
91. Davies, G.F. and F. Pribac, Mesozoic seafloor subsidence and the Darwin Rise, past and present, in *The Mesozoic Pacific*, M. Pringle, *et al.*, Editors. 1993, Washington, DC: American Geophysical Union. p. 39–52.
92. Hill, R.I., Starting plumes and continental breakup. *Earth Planet. Sci. Lett.*, 1991. **104**: p. 398–416.
93. Griffiths, R.W. and I.H. Campbell, Interaction of mantle plume heads with the earth's surface and onset of small-scale convection. *J. Geophys. Res.*, 1991. **96**: p. 18295–18310.
94. Jackson, I., *et al.*, Grain-size-sensitive seismic wave attenuation in polycrystalline olivine. *J. Geophys. Res.*, 2002. **107**: 2360.
95. Forsyth, D. and S. Uyeda, On the relative importance of the driving forces of plate motion. *Geophys. J. R. Astron. Soc.*, 1975. **43**: p. 163–200.
96. Stein, C.A. and S. Stein, A model for the global variation in oceanic depth and heat flow with lithospheric age. *Nature*, 1992. **359**: p. 123–129.
97. Hillier, J.K., Subsidence of 'normal' seafloor: observations do indicate 'flattening'. *J. Geophys. Res.*, 2010. **115**: B03102, doi:10.1029/2008JB005994.
98. Korenaga, T. and J. Korenaga, Subsidence of normal oceanic lithosphere, apparent thermal expansivity, and seafloor flattening. *Earth Planet. Sci. Lett.*, 2008. **268**: p. 41–51.
99. Parsons, B. and J.G. Sclater, An analysis of the variation of ocean floor bathymetry and heat flow with age. *J. Geophys. Res.*, 1977. **82**: p. 803–827.
100. Mooney, W.D., G. Laske, and T.G. Masters, CRUST 5.1: a global crustal model at $5° \times 5°$. *J. Geophys. Res.*, 1998. **103**: p. 727–747.
101. Panasyuk, S.V., Residual topography of the earth. 1998: unpublished.
102. Ishii, M. and J. Tromp, Constraining large-scale mantle heterogeneity using mantle and inner-core sensitive normal modes. *Phys. Earth Planet. Inter.*, 2004. **146**: p. 113–124.
103. Su, W. and A.M. Dziewonski, Simultaneous inversion for 3-D variations in shear and bulk velocity in the mantle. *Phys. Earth Planet. Inter.*, 1997. **100**: p. 135–156.
104. Simmons, N.A., A.M. Forte, and S.P. Grand, Thermochemical structure and dynamics of the African superplume. *Geophys. Res. Lett.*, 2007. **34**: doi:10.1029/2006GL028009.
105. Menard, H.W., Darwin reprise. *J. Geophys. Res.*, 1984. **89**: p. 9960–9968.
106. McKenzie, D.P. and N. Weiss, Speculations on the thermal and tectonic history of the earth. *Geophys. J. R. Astron. Soc.*, 1975. **42**: p. 131–174.
107. Wasserburg, G.J. and D.J. DePaolo, Models of earth structure inferred from neodymium and strontium isotopic abundances. *Proc. Natl. Acad. Sci. USA*, 1979. **76**: p. 3594–3598.
108. Kellogg, L.H., B.H. Hager, and R.D. Van Der Hilst, Compositional stratification in the deep mantle. *Science*, 1999. **283**: p. 1881–1884.

109. Fei, Y., *et al.*, Experimentally determined postspinel transformation boundary in Mg_2SiO_4 using MgO as an internal pressure standard and its geophysical implications. *J. Geophys. Res.*, 2004. **109**: doi:10.1029/2003JB002562.
110. Lay, T., Q. Williams, and E.J. Garnero, The core–mantle boundary layer and deep Earth dynamics. *Nature*, 1998. **392**: p. 461–468.
111. Isacks, B., J. Oliver, and L.R. Sykes, Seismology and the new global tectonics. *J. Geophys. Res.*, 1968. **73**: p. 5855–5899.
112. Davies, G.F., Whole mantle convection and plate tectonics. *Geophys. J. R. Astron. Soc.*, 1977. **49**: p. 459–486.
113. O'Connell, R.J., On the scale of mantle convection. *Tectonophysics*, 1977. **38**: p. 119–136.
114. DePaolo, D.J. and G.J. Wasserburg, Inferences about mantle sources and mantle structure from variations of $^{143}Nd/^{144}Nd$. *Geophys. Res. Lett.*, 1976. **3**: p. 743–746.
115. Davies, G.F., Geophysical and isotopic constraints on mantle convection: an interim synthesis. *J. Geophys. Res.*, 1984. **89**: p. 6017–6040.
116. Widiyantoro, S., Studies of seismic tomography on regional and global scale (PhD Thesis). 1997, Australian National University.
117. Van Der Hilst, R.D. and K. Kárason, Compositional heterogeneity in the bottom 1000 km of Earth's mantle: towards a hybrid convection model. *Science*, 1999. **283**: p. 1885–1888.
118. Davies, G.F., Reconciling the geophysical and geochemical mantles: plume flows, heterogeneities and disequilibrium. *Geochem. Geophys. Geosyst.*, 2009. **10**: doi:10.1029/2009GC002634.
119. McNamara, A.K. and S. Zhong, Thermochemical structures beneath Africa and the Pacific Ocean. *Nature*, 2005. **437**: p. 1136–1139.
120. Hooper, P.R., The timing of crustal extension and the eruption of continental flood basalts. *Nature*, 1990. **345**: p. 246–249.
121. Larson, R.L., Latest pulse of the earth: evidence for a mid-Cretaceous superplume. *Geology*, 1991. **19**: p. 547–550.
122. Christensen, U.R. and A.W. Hofmann, Segregation of subducted oceanic crust in the convecting mantle. *J. Geophys. Res.*, 1994. **99**: p. 19867–19884.
123. Davies, G.F., Controls on density stratification in the early Earth. *Geochem. Geophys. Geosyst.*, 2007. **8**: Q04006, doi:10.1029/2006GC001414.
124. McKenzie, D.P., Some remarks on heat flow and gravity anomalies. *J. Geophys. Res.*, 1967. **72**: p. 6261–6273.
125. McKenzie, D.P., J.M. Roberts, and N.O. Weiss, Convection in the earth's mantle: towards a numerical solution. *J. Fluid Mech.*, 1974. **62**: p. 465–538.
126. Richter, F.M., Convection and the large-scale circulation of the mantle. *J. Geophys. Res.*, 1973. **78**: p. 8735–8745.
127. Parsons, B. and D.P. McKenzie, Mantle convection and the thermal structure of the plates. *J. Geophys. Res.*, 1978. **83**: p. 4485–4496.
128. Yuen, D.A., W.R. Peltier, and G. Schubert, On the existence of a second scale of convection in the upper mantle. *Geophys. J. R. Astron. Soc.*, 1981. **65**: p. 171–190.
129. Sandwell, D.T. and M.L. Renkin, Compensation of swells and plateaus in the north Pacific: no direct evidence for mantle convection. *J. Geophys. Res.*, 1988. **93**: p. 2775–2783.
130. Watts, A.B., *et al.*, The relationship between gravity and bathymetry in the Pacific Ocean. *Geophys. J. R. Astron. Soc.*, 1985. **83**: p. 263–298.
131. Davies, G.F., Ocean bathymetry and mantle convection, 2. Small-scale flow. *J. Geophys. Res.*, 1988. **93**: p. 10481–10488.

132. Haxby, W.F. and J.K. Weissel, Evidence for small-scale mantle convection from Seasat altimeter data. *J. Geophys. Res.*, 1986. **91**: p. 3507–3520.
133. Sandwell, D.T., *et al.*, Evidence for diffuse extension of the Pacific plate from Pukapuka ridges and cross-grain gravity lineations. *J. Geophys. Res.*, 1995. **100**: p. 15087–15099.
134. O'Connell, R.J. and B.H. Hager, On the thermal state of the earth, in *Physics of the Earth's Interior*, A. Dziewonski and E. Boschi, Editors. 1980, Amsterdam: North-Holland. p. 270–317.
135. King, S.D. and D.L. Anderson, Edge-driven convection. *Earth Planet. Sci. Lett.*, 1998. **160**: p. 289–296.
136. King, S.D. and D.L. Anderson, An alternative mechanism of flood basalt formation. *Earth Planet. Sci. Lett.*, 1995. **136**: p. 269–279.
137. King, S.D. and J. Ritsema, African hot spot volcanism: small-scale convection in the upper mantle beneath cratons. *Science*, 2000. **290**: p. 1137–1140.
138. Green, D.H. and T.J. Falloon, Pyrolite: a Ringwood concept and its current expression, in *The Earth's Mantle: Composition, Structure and Evolution*, I.N.S. Jackson, Editor. 1998, Cambridge: Cambridge University Press. p. 311–378.
139. Press, W.H., *et al.*, *Numerical Recipes*. 1986, Cambridge: Cambridge University Press. 818p.
140. McDonough, W.F. and S.-S. Sun, The composition of the Earth. *Chem. Geol.*, 1995. **120**: p. 223–253.
141. Stacey, F.D., *Physics of the Earth*. 3rd edn. 1992, Brisbane: Brookfield Press. 513p.
142. Rudnick, R.L. and D.M. Fountain, Nature and composition of the continental crust: a lower crustal perspective. *Rev. Geophys.*, 1995. **33**: p. 267–309.
143. Labrosse, S. and C. Jaupart, Thermal evolution of the Earth: secular changes and fluctuations of plate characteristics. *Earth Planet. Sci. Lett.*, 2007. **260**: p. 465–481.
144. Cogné, J.-P. and E. Humler, Global scale patterns of continental fragmentation: Wilson's cycles as a constraint for long-term sea-level changes. *Earth Planet. Sci. Lett.*, 2008. **273**: p. 251–259.
145. Korenaga, J., Eustasy, supercontinental insulation, and the temporal variability of terrestrial heat flux. *Earth Planet. Sci. Lett.*, 2007. **257**: p. 350–358.
146. Davies, G.F., Effect of plate bending on the Urey ratio and the thermal evolution of the mantle. *Earth Planet. Sci. Lett.*, 2009. **287**: p. 513–518.
147. Silver, P.G. and M.D. Behn, Intermittent plate tectonics? *Science*, 2008. **319**: p. 85–88.
148. Korenaga, J., Archean geodynamics and the thermal evolution of the Earth, in *Archean Geodynamics and Environments*, K. Benn, J.-C. Mareschal, and K.C. Condie, Editors. 2006, Washington DC: American Geophysical Union. p. 7–32.
149. Korenaga, J., Urey ratio and the structure and evolution of Earth's mantle. *Rev. Geophys.*, 2008. **46**: doi:10.1029/2007RG000241.
150. Wu, B., *et al.*, Reconciling strong slab pull and weak plate bending: the plate motion constraint on the strength of mantle slabs. *Earth Planet. Sci. Lett.*, 2008. **272**: p. 412–421.
151. Capitanio, F.A., G. Morra, and S. Goes, Dynamics of plate bending at the trench and slab–plate coupling. *Geochem. Geophys. Geosyst.*, 2009. **10**: doi:10.1029/2008GC002348.
152. Hirose, K., *et al.*, The fate of subducted basaltic crust in the Earth's lower mantle. *Nature*, 1999. **397**: p. 53–56.
153. Hirose, K., *et al.*, Phase transition and density of subducted MORB crust in the lower mantle. *Earth Planet. Sci. Lett.*, 2005. **237**: p. 239–251.

154. Davies, G.F., On the emergence of plate tectonics. *Geology*, 1992. **20**: p. 963–966.
155. Davies, G.F., Gravitational depletion of the early Earth's upper mantle and the viability of early plate tectonics. *Earth Planet. Sci. Lett.*, 2006. **243**: p. 376–382.
156. Davies, G.F., Episodic layering of the early mantle by the 'basalt barrier' mechanism. *Earth Planet. Sci. Lett.*, 2008. **275**: p. 382–392.
157. Gurnis, M., Large-scale mantle convection and the aggregation and dispersal of supercontinents. *Nature*, 1988. **332**: p. 695–699.
158. Tollo, R.P., *et al.*, Editors, *Proterozoic Tectonic Evolution of the Grenville Orogen in North America*. Memoir Vol. 197. 2004, Boulder, CO: Geological Society of America.
159. O'Neill, C., *et al.*, Episodic Precambrian subduction. *Earth Planet. Sci. Lett.*, 2007. **262**: p. 552–562.
160. Condie, K.C., *Earth as an Evolving Planetary System*. 2004, Amsterdam: Elsevier. 350p.
161. van Kranendonk, M.J., R.H. Smithies, and V.C. Bennett, Editors. *Earth's Oldest Rocks*. Developments in Precambrian Geology, Vol. 15. 2007, Amsterdam: Elsevier.
162. McCulloch, M.T. and V.C. Bennett, Progressive growth of the Earth's continental crust and depleted mantle: geochemical constraints. *Geochim. Cosmochim. Acta*, 1994. **58**: p. 4717–4738.
163. Condie, K.C. and V. Pease, Editors. *When Did Plate Tectonics Begin on Planet Earth?* Special Paper, Vol. 440. 2008, Boulder, CO: Geological Society of America.
164. Mojzsis, S.J., T.M. Harrison, and R.T. Pidgeon, Oxygen-isotope evidence from ancient zircons for liquid water at the Earth's surface 4300 Myr ago. *Nature*, 2001. **409**: p. 178–181.
165. Wilde, S.A., *et al.*, Evidence from detrital zircons for the existence of continental crust and oceans on the Earth 4.4 Gyr ago. *Nature*, 2001. **409**: p. 175–178.
166. Watson, E.B. and T.M. Harrison, Zircon thermometer reveals minimum melting conditions on earliest Earth. *Science*, 2005. **308**: p. 841–844.
167. Campbell, I.H., R. Griffiths, and R.I. Hill, Melting in an Archaean mantle plume: heads it's basalts, tails it's komatiites. *Nature*, 1989. **339**: p. 697–699.
168. Campbell, I.H. and R.W. Griffiths, The changing nature of mantle hotspots through time: implications for the chemical evolution of the mantle. *J. Geol.*, 1992. **92**: p. 497–523.
169. Campbell, I.H., The mantle's chemical structure: insights from the melting products of mantle plumes, in *The Earth's Mantle: Composition, Structure and Evolution*, I.N.S. Jackson, Editor. 1998, Cambridge: Cambridge University Press. p. 259–310.
170. Ernst, R.E. and K.L. Buchan, The use of mafic dike swarms in identifying and locating mantle plumes, in *Mantle Plumes: Their Identification Through Time*, R.E. Ernst and K.L. Buchan, Editors. 2001, Geol. Soc. Am. Special Paper 352. Boulder, CO: Geological Society of America. p. 247–265.
171. Bédard, J., A catalytic delamination-driven model for coupled genesis of Archaean crust and sub-continental lithospheric mantle. *Geochim. Cosmochim. Acta*, 2006. **70**: p. 1188–1214.
172. Smithies, R.H., M.J. van Kranendonk, and D.C. Champion, It started with a plume – early Archaean basaltic proto-continental crust. *Earth Planet. Sci. Lett.*, 2005. **238**: p. 284–297.
173. Bogdanova, S., *et al.*, From Rodinia to Nuna and beyond: Precambrian supercontinent reconstructions delving deeper in time, in *33rd International Geological Congress*, 2008, Oslo. HPP-04. http://33igc.org/coco/entrypage.aspx.

174. Murphy, J.B., et al., Supercontinent reconstruction from recognition of leading continental edges. *Geology*, 2009. **37**: p. 595–598.
175. Davies, G.F., Punctuated tectonic evolution of the earth. *Earth Planet. Sci. Lett.*, 1995. **136**: p. 363–379.
176. Hofmann, A.W., Mantle chemistry: the message from oceanic volcanism. *Nature*, 1997. **385**: p. 219–229.
177. White, W.M., Sources of oceanic basalts: radiogenic isotopes evidence. *Geology*, 1985. **13**: p. 115–118.
178. Zindler, A. and S. Hart, Chemical geodynamics. *Annu. Rev. Earth Planet. Sci.*, 1986. **14**: p. 493–570.
179. Albarède, F., *Introduction to Geochemical Modeling*. 1995, Cambridge: Cambridge University Press. 543p.
180. McDougall, I. and M. Honda, Primordial solar noble-gas component in the earth: consequences for the origin and evolution of the earth and its atmosphere, in *The Earth's Mantle: Composition, Structure and Evolution*, I.N.S. Jackson, Editor. 1998, Cambridge: Cambridge University Press. p. 159–187.
181. Porcelli, D. and G.J. Wasserburg, Mass transfer of helium, neon, argon and xenon through a steady-state upper mantle. *Geochim. Cosmochim. Acta*, 1995. **59**: p. 4921–4937.
182. O'Neill, H.S.C. and H. Palme, Composition of the silicate Earth: implications for accretion and core formation, in *The Earth's Mantle: Composition, Structure and Evolution*, I.N.S. Jackson, Editor. 1998, Cambridge: Cambridge University Press. p. 3–126.
183. McDonough, W.F., Compositional model for the Earth's core, in *Treatise on Geochemistry*, R.W. Carlson, H.D. Holland, and K.K. Turekian, Editors. 2003, Oxford: Elsevier. p. 547–569.
184. Jochum, K.P., et al., K, U and Th in mid-ocean ridge basalt glasses and heat production, K/U and K/Rb in the mantle. *Nature*, 1983. **306**: p. 431–436.
185. Salters, V.J.M. and A. Stracke, Composition of the depleted mantle. *Geochem. Geophys. Geosyst.*, 2004. **5**: 10.1029/2003GC000597.
186. Workman, R.K. and S.R. Hart, Major and trace element composition of the depleted MORB mantle (DMM). *Earth Planet. Sci. Lett.*, 2005. **231**: p. 53–72.
187. Lyubetskaya, T. and J. Korenaga, Chemical composition of Earth's primitive mantle and its variance: 1. Method and results. *J. Geophys. Res.*, 2007. **112**: doi:10.1029/2005JB004223.
188. Sun, S.-S. and W.F. McDonough, Chemical and isotopic characteristics of oceanic basalts: implications for mantle composition and processes, in *Magmatism in Ocean Basins*, Geol. Soc. Spec. Publ. 42, A.D. Saunders and M.J. Norry, Editors. 1988, London: Geological Society of London. p. 313–345.
189. Donnelly, K.E., et al., Origin of enriched ocean ridge basalts and implications for mantle dynamics. *Earth Planet. Sci. Lett.*, 2004. **226**: p. 347–366.
190. Hofmann, A.W., Chemical differentiation of the Earth: the relationship between mantle, continental crust, and oceanic crust. *Earth Planet. Sci. Lett.*, 1988. **90**: p. 297–314.
191. Hofmann, A.W., et al., Nb and Pb in oceanic basalts: new constraints on mantle evolution. *Earth Planet. Sci. Lett.*, 1986. **79**: p. 33–45.
192. Dosso, L., et al., The age and distribution of mantle heterogeneity along the Mid-Atlantic Ridge (31–41 °N). *Earth Planet. Sci. Lett.*, 1999. **170**: p. 269–286.
193. Niu, Y. and R. Batiza, Trace element evidence from seamounts for recycled oceanic crust in the eastern equatorial Pacific mantle. *Geochem. Geophys. Geosyst.*, 1997. **3**: 10.1029/2002GC000250.

194. Zindler, A., H. Staudigel, and R. Batiza, Isotope and trace element geochemistry of young Pacific seamounts: implications for the scale of upper mantle heterogeneity. *Earth Planet. Sci. Lett.*, 1984. **70**: p. 175–195.
195. Allegre, C.J. and D.L. Turcotte, Implications of a two-component marble-cake mantle. *Nature*, 1986. **323**: p. 123–127.
196. Ringwood, A.E., *Composition and Petrology of the Earth's Mantle*. 1975, New York: McGraw-Hill. 618p.
197. Hofmann, A.W. and S.R. Hart, An assessment of local and regional isotopic equilibrium in the mantle. *Earth Planet. Sci. Lett.*, 1978. **38**: p. 4–62.
198. van Keken, P.E. and S. Zhong, Mixing in a 3D spherical model of present day mantle convection. *Earth Planet. Sci. Lett.*, 1999. **171**: p. 533–547.
199. Kellogg, L.H. and D.L. Turcotte, Mixing and the distribution of heterogeneities in a chaotically convecting mantle. *J. Geophys. Res.*, 1990. **95**: p. 421–432.
200. Davies, G.F., Comment on 'Mixing by time-dependent convection' by U. Christensen. *Earth Planet. Sci. Lett.*, 1990. **98**: p. 405–407.
201. van Keken, P.E., E. Hauri, and C.J. Ballentine, Mantle mixing: the generation, preservation and destruction of mantle heterogeneity. *Annu. Rev. Earth Planet. Sci.*, 2002. **30**: p. 493–525.
202. Spandler, C., *et al.*, Phase relations and melting of anhydrous K-bearing eclogites from 1200 to 1600 °C and 3 to 5 GPa. *J. Petrol.*, 2008. **49**: p. 771–795.
203. Davies, G.F., Stirring geochemistry in mantle convection models with stiff plates and slabs. *Geochim. Cosmochim. Acta*, 2002. **66**: p. 3125–3142.
204. Kogiso, T., M.M. Hirschmann, and P.W. Reiners, Length scales of mantle heterogeneities and their relationship to ocean island basalt geochemistry. *Geochim. Cosmochim. Acta*, 2004. **68**: p. 345–360.
205. Sobolev, A.V., *et al.*, The amount of recycled crust in sources of mantle-derived melts. *Science*, 2007. **316**: p. 412–417.
206. Yaxley, G.M. and D.H. Green, Reactions between eclogite and peridotite: mantle refertilisation by subduction of oceanic crust. *Schweiz. Mineral. Petrog. Mitt.*, 1998. **78**: p. 243–255.
207. Pertermann, M. and M.M. Hirschmann, Partial melting experiments on a MORB-like pyroxenite between 2 and 3 GPa: constraints on the presence of pyroxenite in basalt source regions from solidus location and melting rate. *J. Geophys. Res.*, 2003. **108**: doi:10.1029/2000JB000118.
208. Spiegelman, M. and J.R. Reynolds, Combined dynamic and geochemical evidence for convergent melt flow beneath the East Pacific Rise. *Nature*, 1999. **402**: p. 282–285.
209. Sobolev, A.V., *et al.*, An olivine-free mantle source of Hawaiian shield basalts. *Nature*, 2005. **434**: p. 590–597.
210. Takahashi, E., K. Nakajima, and T.L. Wright, Origin of the Columbia River basalts: melting model of a heterogeneous plume head. *Earth Planet. Sci. Lett.*, 1998. **162**: p. 63–80.
211. Salters, V.J.M. and H.J.B. Dick, Mineralogy of the mid-ocean-ridge basalt source from neodymium isotopic composition of abyssal peridotites. *Nature*, 2002. **418**: p. 68–72.
212. Hart, S.R., *et al.*, Mantle plumes and entrainment: isotopic evidence. *Science*, 1992. **256**: p. 517–520.
213. Ito, E. and J.J. Mahoney, Melting a high ^{3}He/^{4}He source in a heterogeneous mantle. *Geochem. Geophys. Geosyst.*, 2006. **7**: doi:10.1029/2005GC001158.
214. Allegre, C.J., T. Staudacher, and P. Sarda, Rare gas systematics: formation of the atmosphere, evolution and structure of the earth's mantle. *Earth. Planet. Sci. Lett.*, 1987. **81**: p. 127–150.

215. Huang, J. and G.F. Davies, Stirring in three-dimensional mantle convection models and its implications for geochemistry: passive tracers. *Geochem. Geophys. Geosyst.*, 2007: Q03017, doi:10.1029/2006GC001312.
216. Huang, J. and G.F. Davies, Stirring in three-dimensional mantle convection models and implications for geochemistry: 2. Heavy tracers. *Geochem. Geophys. Geosyst.*, 2007. **8**: Q07004, doi:10.1029/2007GC001621.
217. Huang, J. and G.F. Davies, Geochemical processing in a three-dimensional regional spherical shell model of mantle convection. *Geochem. Geophys. Geosyst.*, 2007. **8**: doi:10.1029/2007GC001625.
218. Xie, S. and P.J. Tackley, Evolution of U–Pb and Sm–Nd systems in numerical models of mantle convection and plate tectonics. *J. Geophys. Res.*, 2004. **109**: doi:10.1029/2004JB003176.
219. Brandenburg, J.P., et al., A multiple-system study of the geochemical evolution of the mantle with force-balanced plates and thermochemical effects. *Earth Planet. Sci. Lett.*, 2008. **276**: p. 1–13.
220. van Keken, P.E. and C.J. Ballentine, Whole-mantle versus layered-mantle convection and the role of a high-viscosity lower mantle in terrestrial volatile evolution. *Earth Planet. Sci. Lett.*, 1998. **156**: p. 19–32.
221. van Keken, P.E. and C.J. Ballentine, Dynamical models of mantle volatile evolution and the role of phase transitions and temperature-dependent rheology. *J. Geophys. Res.*, 1999. **104**: p. 7137–7151.
222. Chase, C.G., Oceanic island Pb: two-stage histories and mantle evolution. *Earth Planet. Sci. Lett*, 1981. **52**: p. 277–284.
223. Sleep, N.H., Gradual entrainment of a chemical layer at the base of the mantle by overlying convection. *Geophys. J. Int.*, 1988. **95**: p. 437–447.
224. Christensen, U.R., Mixing by time-dependent convection. *Earth Planet. Sci. Lett.*, 1989. **95**: p. 382–394.
225. Allegre, C.J., A. Hofmann, and K. O'Nions, The argon constraints on mantle structure. *Geophys. Res. Lett.*, 1996. **23**: p. 3555–3557.
226. Allegre, C.J., et al., Topology in isotopic multispace and origin of mantle chemical heterogeneities. *Earth Planet. Sci. Lett.*, 1987. **81**: p. 319–337.
227. Davies, G.F., Noble gases in the dynamic mantle. *Geochem. Geophys. Geosyst.*, 2010. **11**: Q03005, doi:10.1029/2009GC002801.
228. Farley, K.A., et al., Constraints on mantle ^{3}He fluxes and deep-sea circulation from an ocean general circulation model. *J. Geophys. Res.*, 1995. **100**: p. 3829–3839.
229. Ballentine, C.J., et al., Numerical models, geochemistry and the zero-paradox noble-gas mantle. *Phil. Trans. R. Soc. Lond. A*, 2002. **360**: p. 2611–2631.
230. Yatsevich, I. and M. Honda, Production of nucleogenic neon in the Earth from natural radioactive decay. *J. Geophys. Res.*, 1997. **102**: p. 10291–10298.
231. Sarda, P., T. Staudacher, and C.J. Allegre, Neon isotopes in submarine basalts. *Earth Planet. Sci. Lett.*, 1988. **91**: p. 73–88.
232. Moreira, M., J. Kunz, and C. Allegre, Rare gas systematics in popping rock: isotopic and elemental compositions in the upper mantle. *Science*, 1998. **279**: p. 1178–1181.
233. Turner, G., The outgassing history of the earth's atmosphere. *J. Geol. Soc. Lond.*, 1989. **146**: p. 147–154.
234. Mahaffy, P.R., et al., Galileo probe measurements of D/H and ^{3}He/^{4}He in Jupiter's atmosphere. *Space Sci. Rev.*, 1998. **84**: p. 251–263.
235. Albarède, F., Time-dependent models of U–Th–He and K–Ar evolution and the layering of mantle convection. *Chem. Geol.*, 1998. **145**: p. 413–429.

236. Lassiter, J.C., Role of recycled oceanic crust in the potassium and argon budget of the Earth: toward a resolution of the 'missing argon' problem. *Geochem. Geophys. Geosyst.*, 2004. **5**: doi:10.1029/2004GC000711.
237. Arevalo, R.J., W.F. McDonough, and M. Luong, The K/U ratio of the silicate Earth: insights into mantle composition, structure and thermal evolution. *Earth Planet. Sci. Lett.*, 2009. **278**: p. 361–369.
238. Taylor, S.R. and S.M. McLennan, The geochemical evolution of the continental crust. *Rev. Geophys.*, 1995. **33**: p. 241–265.
239. Melosh, H.J. and A.M. Vickery, Impact erosion of the primordial atmosphere of Mars. *Nature*, 1989. **338**: p. 487–489.
240. Davies, G.F., Geophysically constrained mantle mass flows and the ^{40}Ar budget: a degassed lower mantle? *Earth Planet. Sci. Lett*, 1999. **166**: p. 149–162.
241. Tolstikhin, I. and A.W. Hofmann, Early crust on top of the Earth's core. *Phys. Earth Planet. Inter.*, 2005. **148**: p. 109–130.
242. Galer, S.J.G., S.L. Goldstein, and R.K. O'Nions, Limits on chemical and convective isolation in the earth's interior. *Chem. Geol.*, 1989. **75**: p. 257–290.
243. McKenzie, D.P. and M.J. Bickle, The volume and composition of melt generated by extension of the lithosphere. *J. Petrol.*, 1988. **29**: p. 625–679.

Index

activation energy, 35
Airy, George B., 26
argon, 158, 193
    budget, 203
aspect ratio, 62
asthenosphere, 8, 27
atmosphere, 155

basalt barrier, 142, 152
basaltic accumulation, 138, 187
boundary layer, 38
    thermal, 38, 46, 47, 51, 59, 62, 71, 77, 102, 104, 150
brittle solid, 18, 57, 63, 71
buoyancy, 39, 41–42, 43, 44, 51, 59, 79
    compositional, 101, 136, 139, 146

Carey, Sam, 14
chemical disequilibrium, 171, 174, 182
Christensen, U., 138, 185, 193
continental collision, 146, 153
continental drift, 13, 17
convection, 38–39, 51
    compositional, 39
    edge, 121
    layered, 112, 142
    mantle, 38, 51, 59, 71
    small-scale, 66, 118
    thermal, 39
core, 4, 77, 83, 103, 127
crust, 4
    buoyancy, 139
    continental, 6, 57, 133, 155, 162
    oceanic, 7
    oceanic thickness, 142
    subducted oceanic, 137, 170

D'' region, 5, 118, 139, 161, 163, 185, 194
Daly, Reginald, 14
decoupling layer, 109
degassing timescale, 199
depletion, 141, 162, 178, 183

Dietz, Robert, 14
drip tectonics, 150
du Toit, Alex, 14

eclogite, 164
    melting, 171
end-members, 180
enriched MORB, 179
enrichment, 156, 163
episode, 144, 149, 151, 152, 153
evolution
    chemical, 154
    mantle, 3, 124
    noble gases, 199
    tectonic, 3, 147
    thermal. *See* thermal evolution

fault
    transcurrent, 16
    transform, 17, 108
fault plane solution, 23
fixed hotspot hypothesis, 75
flood basalt, 2, 95, 116
fracture zone, 14, 16

geochemistry, 3

heat conduction, 3, 45
heat flow, surface, 12, 53, 65, 69, 126
heat loss, 65
heat transport, 63, 64, 66, 81
    plume, 102
Heezen, Bruce, 14
helium, 158, 193
Hess, Harry, 14
heterogeneity, 155, 164–171
    age, 156, 185, 191
    plume, 166
    sources, 165
    survival, 166
Hofmann, A. W., 163
Holmes, Arthur, 14, 18

Index

homogenisation, 192
hotspot track, 74
hotspot, volcanic, 2, 74, 75
hybrid pyroxenite, 174, 176, 194

incompatible element, 176, 194
isostasy, 67
isotope, 154, 156

laminar flow, 166
large igneous province, 152
lead isotopes, 155, 156
lithosphere, 7–8, 27, 36, 56, 60, 71
   bending, 134
   continental, 6, 57
   oceanic, 6, 57
   thickness, 50

magnetic anomaly, 20
magnetic field
   reversal, 19
magnetic stripes, 21
major element, 155, 164, 171
mantle, 4
mantle overturn, 143, 149
mantle plume, 2, 16, 85–88, 101, 127
melt migration, 155, 174
melt recirculation, 175
melting, 155
   heterogeneous, 171, 173
   homogeneous, 172
meteorite, 160
mid-ocean ridge, 9, 60, 71, 107
mid-ocean ridge basalt, 155, 162, 163, 164, 178, 180, 187, 197
mixing line, 176
mobile belt, 17, 58
MORB. *See* mid-ocean ridge basalt
Morgan, W. Jason, 74, 95

neon, 159
noble gases, 3, 155, 193
   abundances, 196, 197
normal MORB, 178
Nusselt number, 54

ocean islands, 15
oceanic island basalt, 155, 156, 163, 180, 181, 187, 193, 201
OIB. *See* oceanic island basalt

palaeomagnetism, 19
Peclet number, 54
peridotite, 164
   composition, 182
phase transformation, 112
plate mode, 63, 69, 71, 105
plate tectonics, 2, 13, 106
   history, 149
   intermittent, 134, 147, 153

plate, tectonic, 17, 51, 58, 71
   thickness, 50
   velocity, 42, 45, 50, 51, 130, 131
plume
   heterogeneity, 181
   mantle. *See* mantle plume
   tail, 86, 91, 102
   tectonics, 105, 149, 152
   thermochemical, 99, 118, 187
plume head, 85, 88, 96, 102
   melting, 98
   velocity, 94
plume mode, 101
polar wander, 14
potassium, 133
primitive lower mantle, 113, 204
primitive mantle, 155, 161, 163, 168, 177
   survival, 170
processing time, 187

radioactive heating, 126, 129, 132
radiogenic heat, 115
Rayleigh number, 52
Rayleigh–Taylor instability, 88
reaction zone, 174
rebound, post-glacial, 31
recirculation, 194
refractory elements, 160, 184
residence time, 186, 187, 189, 193
return flow, 106, 109
rheology, 57, 105, 153
rifting, 106, 116

sampling theory, 190
seafloor flattening, 11, 110, 120
seafloor spreading, 14
seafloor subsidence, 67, 68, 69, 71, 110
seamount, 73, 101
sediments, seafloor, 24
seismic anomaly, 118, 139
seismic tomography, 113
seismology, 22
specific heat, 47
spreading centre. *See* mid-ocean ridge
stirring, 165, 166, 192
   timescale, 166
strain, 29
strain rate, 29
stress, 30, 44
subduction zone, 60
superpile, 112, 118, 194
superplume, 117
swell
   hotspot, 9, 73, 78, 79, 101
   super, 112

tectonic plate. *See* plate, tectonic
thermal conductivity, 46, 48
thermal diffusion, 45
thermal diffusivity, 48, 50
thermal entrainment, 86, 95

thermal evolution
  numerical, 130
  parametrised, 125
thermal expansion, 39, 40, 67
  coefficient of, 40, 41
thermochemical plume. *See* plume:thermochemical
thorium, 133
timescale
  thermal, 49
topography, 8, 66, 108, 119
Tozer, Geoffrey, 36, 130
trace element, 154, 155
  abundance, 176
  incompatible, 155, 162
tracers, 138, 183
transition zone, 5
trenches, ocean, 9
turbulence, 166

undegassed reservoir, 180, 193, 204
uranium, 133, 161
Urey ratio, 126, 132, 133

vertical tectonics, 153
viscosity, 27–30, 43, 44
  mantle, 30, 34
  temperature dependence, 35, 56, 61
viscous fluid, 18, 28, 63, 71
viscous fluid flow, 3, 91
volatile elements, 160
volcanic hotspot. *See* hotspot, volcanic

Wadati–Benioff zone, 112
Wegener, Alfred, 13
wetspot, 121
White, W. M., 163
Wilson, J. Tuzo, 15, 58, 73